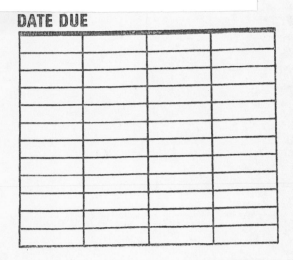

The Inner I

The Inner I

British Literary Autobiography
of the Twentieth Century

BRIAN FINNEY

Oxford University Press · New York
1985

ISBN 0–19–503738–3
Printing (last digit): 9 8 7 6 5 4 3 2 1
Printed in Great Britain

Contents

Acknowledgements

I am indebted to the British Academy for an award from their Small Grants Research Fund which enabled me to undertake research of primary materials for this book in fourteen libraries in the United States. I would like to record my gratitude to the University of London for freeing me from my normal obligations throughout the summer of 1983 when I wrote the majority of this book. I am also grateful to Janet Law and Carolyn Wilde for their detailed comments on an earlier draft of the book.

Finally I would like to acknowledge my thanks to all those publishers listed in the notes for permission to quote extracts from books under their imprint.

Introduction

In view of the widespread popularity of autobiography and of the number of good writers who have turned their hand to it, it has been surprisingly neglected by the English literary establishment. There is hardly a university in the United Kingdom offering it as a subject option. The only standard (and now dated) book on English autobiography as such is by an American.[1] The situation in the United States is very different. American scholars like James Olney and the late Elizabeth Bruss have realized the extent to which autobiography lends itself to the generic and semiotic theories of the Structuralists and Post-Structuralists. Accordingly they have raised the level at which the subject is discussed to a much more interesting and illuminating one. Autobiography is offered as an option in a large number of colleges in the States. The great majority of the critical literature on it is written by Americans. On looking over this considerable body of criticism I was struck by the extent to which it reflected an unconscious bias originating from its origins in the United States. There is an understandable desire among American scholars to trace the development of American autobiography after it became independent of its European roots. The result is a proliferation of books about early European and more recent American autobiography, but a dearth of books on English autobiography of the last two centuries. In fact I was astonished to discover that there was not a single book devoted to English autobiography of the twentieth century. This book is an ambitious attempt to make good this omission and to map out this rich unexplored territory for the benefit of teachers, students and the large number of general readers who account for the popularity of autobiography in modern times.

Obviously it is impossible to do any kind of comprehensive

justice to such a prolific branch of literature. So in the first place I
have confined the autobiographies considered here geographically
to those written by British writers, although I have made excep-
tions of George Moore, W. B. Yeats and Forrest Reid whom at
any rate I see as Anglo-Irish writers. Secondly I have chosen to
concentrate on literary autobiography, that is to say autobiographies written by novelists, poets and playwrights. I have
done this partly because a large majority, although not all, of the
best-written autobiographies are those of practising writers. That
this is the case points to the fact that successful autobiographies
depend more on their literary quality than on the nature of the
lives they are describing. Many imaginative writers have led
reasonably uneventful lives. Their adventures are mainly those of
the mind and the majority of these have usually found their way
indirectly into their imaginative work. But that still leaves them
with their skill at using language for narrative ends. Ultimately
what interests the reader is not so much what story is being told
but how well it is told.

Nevertheless there are many writers who believe, like Bernard
Shaw, that the passive nature of their lives disqualifies them from
even attempting to write their autobiography: 'Things have not
happened to me: on the contrary it is I who have happened to
them; and all my happenings have taken the form of books and
plays. Read them, or spectate them; and you have my whole
story.'[2] The fallacy in Shaw's argument lies in believing that it is
ever possible to tell one's 'whole story'. There are endless ways of
telling one's story, as I aim to show in this book, but none of them
can tell the whole story. Pamela Hansford Johnson, who is apparently aware of this when she argues that 'No novelist should
attempt to write his *full* autobiography',[3] still assumes in focusing
on things and people that have been important to her in her life
that there can be only one version of this aspect of her complete
story. Yet since Barthes at least we are aware of how the text
assumes a life of its own, each time creating a new story which
owes as much to the conventions of the genre as to the life and
opinions of the author. Besides, the act of writing an autobiography can become part of the life being written about, as Montaigne
realized centuries ago in remarking on his autobiographical *Essays*
'I have no more made my book than my book has made me.'[4]

Necessarily this means that each new attempt at autobiography will tell a different story since the story has changed in the course of its telling and as a result of it.

The recent emphasis on the importance of *graphos*, the act of writing, to an understanding of that composite artefact, auto-biography, has led me further to confine my consideration to autobiography as a literary form. Arguments continue to rage over definitions of what is understood by the term 'autobiography'. Some American critics insist that it can refer to not just memoirs, reminiscences, confessions, apologies and the like, but to diaries, letters, novels and even to a poem such as *Four Quartets*. One such critic, William Spengemann, deliberately renders the term meaning-less by an ingenious definition that negates itself: 'Without a self, one cannot write about it, but whatever one writes will be about the self it constructs. Autobiography thus becomes synonymous with symbolic action in any form, and the word ceases to designate a particular kind of writing.'[5] Critics of autobiography have been playing this game for a long time now. Back in the 1930s E. Stuart Bates asserted that 'from a librarian's point of view, an autobiogra-phy is a book which believes itself to be one', and therefore that it 'is not so much a species of literature as an idea'.[6] I don't intend to become entangled with this speculative question of definition and would refer the reader to Elizabeth Bruss's admirably lucid attempt to define the genre at length in the Introduction to her study, *Autobiographical Acts*. I am prepared to accept for present purposes the brief and pithy definition that Motlu Konuk Blasing offers: ' "Autobiographical" refers to works in which the hero, narrator, and author can be identified by the same name',[7] (although I would prefer 'protagonist' to 'hero', and would wish further to define 'works' so as to exclude collections of letters or diary extracts.)

In choosing to restrict myself to autobiography as a literary form I am not necessarily making a value judgement as to its superiority to other kinds of autobiography. All those memoirs of famous politicians, statesmen, generals, dictators and the like have a legitimate place in the spectrum of autobiography; they are important personal contributions to social, political and military history. There is also the reminiscence, which like the memoir concentrates on the world outside the self, though normally a less

public world than is generally the case with the memoir. Reminiscences provide invaluable biographical information about individuals and social groupings; they resemble group portraits in
which the artist includes himself. They are a sub-genre in their own
right and have their own rewards. Richard Butler coined the term
'allobiography' (from the Greek *allos* meaning 'other') for all such
books that are 'not self-centred'.[8] As opposed to allobiography,
what I understand by 'autobiography' for the purposes of this book
is subjective autobiography, that in which, as Roy Pascal suggests,
attention is focused on the self.[9] This is a peculiarly twentieth-
century form since the focus that the Renaissance initially turned on
individuality had become blurred by Victorian times. Freud and his
successors dramatically reversed this trend. They transformed our
understanding of the self by their exploration of the unconscious
and gave a new impetus to subjective autobiography.

By concentrating on those subjective autobiographies to which
imaginative writers have given literary form I am necessarily
drawing attention to what might seem to be a paradox intrinsic to
the genre. How is it possible for an autobiographer to explore with
honesty the recesses of the self while simultaneously he is trying to
satisfy aesthetic criteria concerning form, structure, tone and the
like? The first section of the book, *Versions of the Truth*, sets out to
demonstrate the extent to which autobiographical narrative, like
other forms of narrative, makes heavy use of aesthetic criteria even
when the writer is most concerned to give an honest 'unvarnished'
picture of the facts. The four chapters in this section form a
continuous argument; they move from the most factual to the most
sophisticatedly artistic examples of autobiography.

The second longer section, *In Search of Identity*, builds on the
assumption formed in the first section that any form of written
self-analysis is bound to assume many of the formal characteristics
of a written narrative, imaginative or otherwise. The twentieth-
century discovery of the numerous ways in which the unconscious
works within us has had revolutionary repercussions for autobiography as it has had for the novel and other artistic genres. In the
second section I look at some of the principal ways in which
modern autobiographers have chosen to portray themselves in the
light of this changed perception of themselves. Two chapters on
childhood and parents reflect the findings of psychoanalysis which

has stressed the way in which early influences and experience are of enormous importance in the formation of the adult self. There is a chapter on the use of autobiography as a therapeutic tool for coming to terms with the self. Another chapter concentrates on dreams and myths which offer access to the unconscious. Throughout these chapters I have chosen primarily to refer, though sparingly, to the theories of Freud and Jung. It seems to me particularly appropriate to apply the theories of these two innovators to modern autobiographies, so many of which are heavily indebted to them for their entire outlook and presuppositions. At the same time I have deliberately concentrated on these two in order to draw attention to the relative nature of the insights which psychoanalysis has to offer. I do not wish to present the theories of Freud and Jung as embodying absolute truths. Rather they offer modern writers – and readers – partial truths relative to their time. The final two chapters widen the focus so that psychoanalytically based insights are placed within a broader view of the self which includes spiritual vision and the interaction of the self with his society and time.

Since I have used both 'man' and 'his' in the paragraph above to stand for both sexes, this is an appropriate point at which to offer my excuses for adopting the conventional use of the masculine pronoun to stand for male and female autobiographers. I was obliged for clarity's sake to employ either the masculine or feminine pronoun throughout. I chose the former because the majority of autobiographers cited in this book are men.

While I am making excuses I had better explain and defend the choice of texts selected for discussion in this book. There is no way I could possibly give attention to all British literary autobiographies published this century. Selection is inevitable. At the same time I am anxious to introduce readers to as many as possible of what I consider to be the best instances of this sub-genre. So I have chosen to concentrate on just under thirty autobiographies, some of more than one volume in length. But I then always made use of them purely as examples for my evolving argument and have never attempted to write comprehensively about every aspect of any of them. Many of the texts chosen could have been used in different chapters for quite different illustrative purposes. Yeats's *Reveries over Childhood and Youth*, for instance, which appears in the chapter on *Parents and Children*, could just as easily have been used to illustrate themes discussed in the chapters on *Childhood*, *Autobiography as Self-Analysis*, *The Spiritual History of the Self*, or *The Double Perspective*. In each case I would have had different things to say about it. Had I had the space I would have added a further chapter on 'Sexual Identity'[10] to the second section of the book. Other possible chapters hover in the corners of my mind. There is no way one can even hope to cover every important aspect of a subject so wide in the space of one book. So I have sought a middle ground between the extremes of an exhaustive analysis of a very small number of autobiographical texts and of a general if superficial survey of this immensely rich field. The shape of the book is controlled by the evolving argument. Its principal texts constitute what I consider to be outstanding contributions to the genre. At the same time I make extensive use of other autobiographies to give support to the main argument; simultaneously they act as reminders of the wider context from which the selection of primary works has been made.

I can still recollect the surprise I felt at the seeming ease with which I sat down one day in March 1983 after five years of preparation and produced an outline of the book from which I have since hardly deviated. André Maurois has argued that all biography is to some extent 'autobiography disguised as biography'.[11] I have become similarly aware of the extent to which my own autobiography lurks concealed behind the argument and approach of this book. The first four chapters retrace my realization, as a teacher and reader, of the limitations of the so-called New Criticism in which I was brought up and which still dominates the teaching of literature in the world of further education to which I belong. I need hardly point out that in following this trend I was simply moving in a similar direction to most practitioners in the field. The switch to a psychoanalytical perspective in the second section virtually co-incides with my own experience of analysis which has in turn affected my previous assumptions about literature. Where previously I had tried hard to subordinate my own subjective responses to texts so as to capture more of the writer's intended effect, I now feel that the only healthy response is openly to relate the text to my own experience of life. This has meant rejecting the allegedly objective approaches claimed for themselves by both the New

Criticism and by Structuralists and their successors. It is only by discovering some correspondence, however remote, between what I read and my own experience of life that I can bring words on the page back to life. I offer no excuses for this personal bias. As I have come to realize in the course of writing this book it is only fitting that I should have adopted an autobiographically motivated and essentially subjective approach to a study of subjective autobiography.

Versions of the Truth

1

Factual Accounts of the Self

What does the average reader understand is meant when a book is categorized as an autobiography? That writers of such books are explicitly writing about their own lives. But more than that. That such writers are attempting to give as truthful an account of themselves as they can manage. This mutual understanding between autobiographers and readers is sometimes the only means of distinguishing between autobiographies and novels written in the first person. This distinction may be embodied in the title alone, or even in such extra-textual material as a foreword. By calling a book an autobiography a writer creates generic expectations in the mind of the reader which are different in kind from those produced by calling a book a novel or a history. Autobiography shares with the novel its interest in individual human nature, but it does not allow itself the novelist's freedom of invention. Like a work of history it tries to be factually true; yet it differs from history in not offering generalized truths based on documentary evidence. Autobiographical truth is concerned with both fact and the meaning the autobiographer attaches to the facts.

What autobiographers understand by 'truthful' is another matter. Autobiographers appear to have as many different conceptions of what constitutes the truth about themselves as readers have different expectations of them. Moreover what is seen as the truth by one age is not necessarily so for another. Augustine and his seventeenth-century successors conceived of the self as the soul. For them any 'truthful' account of the self's or soul's history was inevitably religious, specifically Christian. Even then, as Dean Ebner has shown,[1] Baptists, Anglicans, Quakers and Presbyterians each had quite distinguishable preconceptions about how the soul truly relates to God. On the opening page of his *Confessions*

Rousseau claims that he has never put down as true what he knew to be false. But what Rousseau meant by 'true' was 'true to nature' which constituted a radical break with the Augustinian tradition. The Victorians in their turn retreated from what they saw as this defiantly individualistic definition of the truth. They preferred to focus on a historically factual and publically shared conception of the self. This duly led to a strong reaction in the twentieth century against the hypocrisies this approach produced. Freud's postulation of the unconscious removed what he saw as the inaccessible part of the human psyche from any possibility of direct observation. Only through the interpretation of dreams or of neurotic symptoms could the outlines of the unconscious self be deduced. So, quite apart from the age-old philosophical doubts about whether or how one can perceive oneself, the present century was faced at its inception with the possibility that parts of the self are by their nature unknowable, or at most only partially and indirectly knowable.

It is not surprising, then, to find many modern autobiographers assuming that it is impossible to write a complete history of the self. 'No one can tell the whole truth about himself,' Somerset Maugham insists,[2] if only because, as Tristam Shandy discovers, writing about it takes longer than living it. But it is not simply the process of selection that can distort the truth. The fact is, as André Maurois argues, that 'the severest autobiography remains a piece of special pleading'.[3] The unconscious choices exercised by memory are themselves the product of the earlier workings of an anxiety-ridden ego busy repressing unacceptable elements from the past history of the self. This additional psychological insight into the individual's dependence on memorial distortion and censorship for a sense of self has led many twentieth-century writers to extremes of distrust in their own powers of recollection. George Orwell went so far as to claim that 'autobiography is only to be trusted when it reveals something disgraceful'.[4] Other writers have simply resigned themselves to what Freud saw as the mendacity implicit in the genre.[5] 'All autobiographies are lies,'[6] Shaw flatly asserted.

Many of the most thoughtful twentieth-century autobiographers have consequently taken it for granted that since truth concerning the self is partial and has many determinations all that they can hope to do is to make explicit to the reader what aspects of themselves they are seeking to reveal within their work. Some, like Gerald Brenan, specifically reject the use of such documentary aids to memory as letters on the grounds that 'memory is the ordering and sifting principle'.[7] But, as W. H. Hudson pointed out in his autobiography, 'It is easy to fall into the delusion that the few things . . . distinctly remembered and visualized are precisely those that were most important in our life.'[8] So others, perhaps influenced by the implications in the title of Goethe's autobiography, *Poetry and Truth*, like him give priority to the first term over the second. Yeats, for example, is quite prepared to deny the tyrannical power of facts where these bar entry to his imaginative life. Enlarge the term from poetic to spiritual truth and you find autobiographers as different as W. H. Hudson and John Cowper Powys making similar assumptions to Yeats. Yet others take their hint from Jung and set out to tell their personal myths, deliberately focusing on the constructs of a subjective interior self which for them offers the most valuable truth, however elusive, of the many facets that the self offers for inspection and analysis.

What all these approaches have in common is a recognition that autobiography specifically presents the writer with an opportunity to pursue the truth about himself from within the self. Abandoning any attempt to confine themselves to historically verifiable facts, such writers exploit the realization that the meanings they attach to events in their lives reveal as much or more about themselves as do the facts in isolation. In any case it is not possible to create a living portrait by the mere accumulation of facts. As Georges Gusdorf pointed out back in 1956, 'An autobiography cannot be a pure and simple record of existence, an account book or a log book' because 'a record of this kind, no matter how minutely exact, would be no more than a caricature of real life.'[9] Besides, the juxtaposition of facts becomes in itself an act of interpretation. However valuable such books are as source material, they fail as works of literature. Their authors seek to deny the interpretative element, the meanings that individuals attach to their lives and which normally constitute the structuring principle of their autobiographies. They offer an impoverished version of the self as an object that acts and speaks, is acted upon and spoken to, but which has no inner life, no psychic drama which can at times usurp the life of the self that interacts with the world around it. Their concept of autobiographical truth is restrictive and their trust in memory naive. As a result they deprive themselves of that fascinating interplay between inescapable biographical fact and imaginative interpretation, even reconstruction, which imbues autobiography at its best with the dual satisfaction of historical veracity and artistic creativity.

There is obviously a place for factual records of a life within the breadth of the autobiographical spectrum. In particular they can be an invaluable minefield of information for readers interested in their sphere of activity. But as works of literature that aim at self-portraiture they normally fail to satisfy because they abjure that self-regarding, self-interpretative element which lies at the heart of the genre whenever it is exploited to its fullest. Frequently they are the result of hesitation on the part of writers to engage with the full potential of the genre, either because they are afraid what they will reveal in the process, or out of a misguided sense of what form of truth is proper to the genre. One can think of numerous examples of the former kind who state that they do not wish to hurt the living nor their survivors. Among the latter can be found not just writers of *res gestae*, but creative writers who choose to write about those truths which give meaning to their life in their imaginative work and see their autobiographies primarily as a means of putting the record straight. John Wain's mid-life autobiography, *Sprightly Running*, is an instance in point. In the penultimate chapter he opts out of describing his mixed, complicated reactions to a month's visit he made in 1960 to the Soviet Union as a result of which he claims that he was given a criminal record there. He excuses himself on the grounds that 'such complex material needs to be picked up in the powerful tongs of art, and a book like this isn't art, only the bald record of certain experiences, inner and outer, that make up the raw material of art'.[10] It is especially ironical that Wain's very rejection of art is expressed by his artistic use of metaphor. Good autobiography cannot do without art. Art involves artifice. But interpretation can be as true or false as the facts that are being interpreted. Subjective autobiography at its best cultivates both forms of truth.

W. H. Davies: *The Autobiography of a Super-Tramp*

The problems inherent in a predominantly factual approach are illustrated particularly clearly in the case of W. H. Davies's earliest book of autobiography, *The Autobiography of a Super-Tramp* (1908). His first published work in prose, the book has justifiably earned a reputation for its vivid description of a tramp's life-style, something that few people have experienced or know anything much about. The *Autobiography* covers his first thirty-four years, from his birth in a pub in Newport in 1871 and his upbringing by his paternal grandparents, to his six years of tramping in the United States which was brought to a sudden end by a train accident in 1899 leading to the amputation of a leg. The last third of the book provides a graphic description of the six years he spent in English dosshouses while he was serving his literary apprenticeship and ends with his eventual recognition as a poet after his first book of poems had been published. Because I wish to concentrate on the problems associated with Davies's factual approach to autobiography there is a danger that I may give the impression that I am interested only in carping at his book. In fact it has strengths of its own, especially those relating to plot narration. I am forced for reasons of space to take its strengths for granted. What I am concerned to show here is how it nevertheless limits its achievement by its failure to recognize the full opportunities which the genre offers in the form of interpretative, in addition to factual, truth.

What contributes to this work's status as a classic case history is the ingenuous attitude Davies appears to adopt towards the task of writing about his unusual life. By assuming the role of a 'natural' he denies himself the possibility of using the more sophisticated self-reflective responses present in the genre's potential. Like Wain he explicitly relegates his autobiography to an inferior position to that in which he holds his imaginative writing. In his 'Author's Note to the 1920 Edition' he states: 'My life has been mainly concerned with my poetry, and I had no more ambition in my prose at that time than I have now.'[11] It is a fact that throughout his life Davies wrote his works in prose either at publishers' suggestions or to make money. That did not prevent him from treating his prose work seriously or with artistic integrity. Indeed in the same 'Author's Note' he comments on how much his *Autobiography* owes its popularity to 'a single straight-forward style', together with the 'spirit of adventure' of its contents (p. 15).

What is of interest is that both effects were as much the product of calculation as of spontaneity. Davies's true naivety lay in his failure to understand the extent to which the process of writing transmuted the raw material of life into a verbal artefact. For a start there was the question of selecting from the totality a book's worth of adventures. His first version of the autobiography, written in six weeks, told little of his American or English tramping experiences. Instead it strung together a series of his more sensational moments in dosshouses, slums and brothels on both sides of the Atlantic. He was persuaded to discard these lurid incidents from his past by the combined efforts of Edward Thomas and Edward Garnett, and to concentrate instead on his life spent tramping the States and peddling in England. Many of the episodes relating his sexual encounters were subsequently incorporated in a subsequent volume of prose reminiscences, *The True Traveller* (1912), although most of them were omitted from *The Adventures of Johnny Walker, Tramp* (1926), a book which sought to turn his autobiographical essays collected in *Beggars* (1909) and *The True Traveller* into a continuous narrative with the introduction of some new material. It is clear, then, that when he first attempted to tell the truth about his past life he attached considerable importance to his sexual life. Yet the revised version of the *Autobiography* omits all mention of it. A rather similar pattern repeated itself when he came to write his second inferior volume of autobiography, *Later Days* (1925), which omitted to describe the extraordinary circumstances under which he came to meet his wife, Emma. This was followed by *Young Emma*, the frank and full story of how he set out to look for a wife whom he eventually picked up at a bus stop one evening. In this case the direct cause of his self-censorship was the 24-year-old Emma who objected to publication, which was consequently delayed until 1980, the year after her death.

There is nothing wrong with reticence just as there is nothing wrong with selectivity *per se*. But Davies is either showing an unusual lack of self-awareness or he is consciously misleading the reader when he claims in the *Autobiography*: 'I am now giving my experience honestly and truthfully, and thought for thought, if not word for word, as they happened' (p. 238). In the final paragraph of the book he further appeals to all his tramping acquaintances to authenticate his experience: 'I have the one consolation to know that many a poor man . . . knows what I have written to be the truth' (p. 243). Maybe. But not the whole truth. Is not this insistence on the truthfulness of his narrative despite important omissions motivated by his need to recreate his earlier image in a form more in keeping with his later celebrity? This suspicion is especially aroused in *Later Days*. There, for example, he describes his coming together with his new wife as happening 'with as much ease as two drops of rain on a leaf when the wind shakes it' (p. 215). Yet we know from *Young Emma* how traumatic his early months with her were. At the time he was seemingly unaware that she was pregnant by another man until she gave premature birth to a stillborn child, and then assumed – incorrectly – that he had caught VD from her. *Young Emma* is remarkably frank, perhaps a retrospective attempt to correct the misleadingly romantic version of their courtship which he had just published in *Later Days*. Whatever conclusions one reaches, the disparity between the two accounts alerts one to the very different kinds of autobiographical truth of which Davies avails himself on different occasions.

As one would expect from any autobiography written from memory without the help of a diary, Davies's book has its quota of inaccuracies, as well as omissions. What is surprising about the mistakes he makes is the way he seems to get wrong many of the facts concerning the most significant events in his life up to his thirty-fourth year. He gets his date of birth wrong (3 July, not 20 April). In *The Adventures of Johnny Walker, Tramp* he denies meeting Brum, the beggar, until after his first stay in Chicago, whereas in the *Autobiography* it is Brum, his instructor in the arts of tramping and begging, who first shows him how to beat his way by train to Chicago. Davies even gets wrong the date and circumstances of his train accident at Renfrew in which he lost his right leg. It occurred on 20 March (whereas he claims in the *Autobiography* (p. 129) to have left London less than two weeks earlier, an impossibility), and his foot was crushed, not severed as he asserts.[12] The errors are nevertheless understandable when one considers the effect trauma of this kind was on memory. Equally important to his subsequent development as a poet was his decision, after reading an appreciation of Burns, to turn poet himself. In the *Autobiography* he dates this turning point in his life as happening in Chicago in October 1898 when he decided to return to England, only to fritter his money away once there without any attempt at writing poetry. Instead he is lured to the Klondike in Canada by an article in a London evening paper. It is on his way to the Klondike that he slips and crushes his leg under a train at Renfrew. But in later life, according to his biographer, Richard Stonesifer, he told his friends that it was during his six weeks' recovery in Renfrew hospital that he read the article on Burns and decided to turn poet.[13]

There is nothing unusual or intrinsically blameworthy in the occurrence of these inaccuracies. What is to the point is that Davies fails to allow for their occurrence. The attitude he assumes throughout is that of a 'natural', someone so transparently without deceit that the reader is lured into uncritical acceptance of the truth of all he says. Take, for example, his confession that he wrote a letter home on the first night of his initial boat trip across the Atlantic in which he announced his arrival in America. Its contents were based entirely upon a description of the States by a man who had been there and whom he had met in Liverpool. Davies continues:

This letter was given to the steward at Queenstown, and was written to save me the trouble of writing on my arrival, so that I might have more time to enjoy myself. Several years elapsed before it occurred to me how foolish and thoughtless I had been. The postmark itself would prove that I had not landed in America, and they would also receive the letter several days before it would be due from those distant shores. I can certainly not boast a large amount of common sense (p. 33).

One's first reaction to this passage is to respond to the honesty and simplicity of his confession with admiration. This has certainly been the response of most critics who repeatedly have applied epithets like 'unaffected', 'candid', and 'unvarnished'[14] to his *Autobiography*. But further reflection leads to the realization that Davies is as much put out by his failure to lie convincingly as he is by the pain he might have caused. Next one notices his change from past to present tense: he *had been* foolish; he *is* lacking in common sense. The implication is that the naivety he employs in narrating his life history is no longer of the same order of naivety that led him to write that letter home. Finally one finds oneself asking whether a man who in his youth was capable of practising literary deception on his family out of a need to construct a fantasy account of himself is not still under that compulsion twelve years later when he came to write the *Autobiography*.

The more one ponders over Davies's adoption of an ingenuous mask the less ingenuous he appears. Not that he wears the mask only when he is exposing himself to the gaze of the world. He appears to need to view his own early life through the eyes of an *ingénu* at least partly to exculpate himself in his own eyes. Ezra Pound has talked of Davies's 'peasant shrewdness',[15] a phrase which accurately encapsulates the combination of childlike simplicity and adult calculation which characterizes his *Autobiography*. Peasant shrewdness depends for its effectiveness on disarming the world with a peasant's apparent innocence in order to have one's own way. Half instinct, half pose, it reflects the confusion Davies experienced within himself. In his life and in his account of it he showed both a child's unthinking innocence and an adult's defensive assumption of the cloak of childhood innocence. He was never quite sure which role he was playing at any time, least of all in the *Autobiography*. Yet the self-reflective nature of the genre continually confronted him with the ambiguity underlying his actions, an ambiguity he was seeking to evade.

There is a consequent confusion in the way Davies handles many of the major themes in the book—his delinquent behaviour as a schoolboy, his bouts of alcoholism, his fear of work. His attitude to money and begging is especially ambivalent. On the one hand he constantly castigates himself for his idleness. On the other hand he blames it on the small fortune awaiting him at home. He refers to other beggars as 'scum' (p. 80), yet begs himself. Later in the book he bemoans the lot of the tramp and considers himself 'entitled to a little rest' (p. 182) despite repeated earlier admonitions against the pointlessness of his tramping. Again and again he recounts episodes in which drink gets the better of him, he loses his small savings, and reflects for a sentence or two on his weakness—only to hasten on with the story. These reflective interventions on the part of the narrator of the autobiography deprive his adventures of much of their meaningfulness. Equally the fact that he continually repeats actions which the narrator condemns undermines the sincerity of the narrator's judgements. This is largely the consequence of a failure on Davies's part sufficiently to distinguish between narrator and protagonist in the book. He is completely at ease only when this distinction is submerged because he is pursuing his plot, treating it very much in the 'spirit of adventure' that he mentions in the Author's Note (p. 15).

This concentration on external circumstances and events is another aspect of his 'peasant shrewdness', since it shifts attention from the really difficult, unpleasant or painful questions arising from his actions. Gaoled in New Haven for begging, Davies obtusely objects to the judge's description of him and his companion as 'two brawny scoundrels who would not work' (p. 46): 'However true this might be as applied to us in a moral sense, it certainly was not a literal fact, for we were both small men,' he quips, and continues: 'People who, not seeing us, would read this remark in the local paper, would be misled as to our personal appearance' (p. 46). Clearly Davies is not interested in asking himself why he had such a strong aversion to work. Like Defoe he is excited only by narrating 'literal fact', the unusual things that happened to him in his life outside the social pale. Normally he contents himself with some trite moral comments before moving on to the next episode. There are two kinds of truth to be found in autobiography – truths of fact and truths of interpretation of the facts or truths of meaning. Davies attaches value mainly to the former. When it comes to his feelings about the events in his life he becomes either confused or uninterested. The confusion arises from the difference in feelings that exists between his past and present selves. His constant moral interjections lack the liveliness that characterizes his factual accounts of his past, as if he preferred the unthinking tramp and beggar that he was to the didactic story-teller that he has become.

The style and language that he employs for telling his story is in itself revealing. In his Author's Note of 1920 he claims to have won the respect of the literary world for his use in the book of a 'simple straightforward style'. In fact the style he employs is less simple and more studied than he suggests. Early on in the autobiography he describes how as a boy he developed a habit of waiting before speaking until he could rephrase his thoughts in an acceptable form:

> . . . having heard so much slang my thoughts often clothe themselves in that stuff from their first nakedness. That being the case, shame and confusion in good company make me take so long to undress and clothe them better, in more seemly garments, that other people grow tired of waiting and take upon themselves the honour.

So even in his everyday speech he is carefully formulating his language so as to give it a more 'seemly' appearance. The imagery he employs of clothing his nakedness suggests a deep desire to hide his private self from the world. Words are for him a garment one puts on for purposes of social acceptance. In the passage immediately succeeding the above Davies refers to his first written composition. There he attributes to the act of writing the additional benefit of helping him to 'have successfully navigated the deeps of misery' (p. 30) and to resist the evils of drink. Language, then, both spoken and written, offers a disguise to the world and an escape from the shadow self. To suggest to the reader that he employs a 'simple straightforward style' is but another instance of his use of language as protective clothing.

Although the title, *The Autobiography of a Super-Tramp*, was concocted by Shaw, Davies's ready acceptance of it is one small indication of how ambivalent was his attitude, both psychological and literary, to his past. If one looks at the passages of direct speech used by Davies and his fellow tramps it is clear that he saw himself and them as 'gentlemen of the road'[16] in a quite literal sense. Their manner of expressing themselves is nearly always more restrained and dignified than the officials of society with whom they come into conflict. Take, for example, his first illegal ride on a train to Chicago with Brum, his fellow tramp, when the brakesman discovers them:

'Hallo, where are you going?'
'To the hop fields,' I answered.
'Well,' he sneered, 'I guess you won't get to them on this train, so jump off, at once. Jump! d'ye hear?' he cried, using a great oath, as he saw we were little inclined to obey. Brum was now wide awake. 'If you don't jump at once,' shouted the irate brakesman, 'you will be thrown off.'
'To jump', said Brum quietly, 'will be sure death, and to be thrown off will mean no more' (p. 50).

Brum's calm dignity and courage is contrasted to the course, peremptory and inhuman shouts of the brakesman. The same contrast in speech patterns occurs in the dialogue between the mercenary Judge Stevens ('Pass over the dollars, boys' (p. 56)) and Davies who refuses to pay or to conceal his thoughts when he 'saw real injustice or hypocrisy'at this 'mock trial'. Hypocrisy is also the hallmark of the Charity Organisation whose spokesman arrogantly asks Davies: 'Do you ever do anything for a living?' and, ignoring his wooden leg, adds insult to injury with: 'Is there anything the matter with you?' (p. 230). After the Organisation's spokesman has refused him help for the second time, Davies's parting shot has all the barbed poise of retrospective wish–fulfilment when he tells him, 'I had not come there with any great hopes of receiving benefit, and ... I was not leaving greatly disappointed at this result' (p. 230).

Conversations between tramps in the book normally take the form of mannered gentlemanly exchanges between cultivated and collected individuals. 'Are you going to do business on the road?' asks one of his old companions, meaning, does he still need to beg sufficient money for his night's food and lodging (p. 212). Or listen to Brum proposing that they get themselves arrested in Michigan so as to pass the coldest months of the winter free in the warmth and comfort of gaol:

'There is nothing else but begging before you, for the coming winter,' said Brum, warming to his subject, 'but if you like to enter with me the blessed havens of rest, where one can play cards, smoke or read the time away, you will become strong and ready for work when the spring of the year arrives' (p. 59).

Davies's euphuistic use of language ('do business', 'havens of rest') is further evidence of the art underlying his apparent 'simplicity', an art which derives from the literary, non-realist tradition of the pastoral idyll.

Only rarely are we given glimpses of treachery, meanness and foul language on the part of the tramping community. Davies's evasive use of language reflects a deeper evasion of the consequences of his own actions. When he refers to his addiction to the bottle he employs a metaphorical figure of speech that has the effect of shifting responsibility from himself to 'Drink, my first officer, who many a day and many a night endeavoured to founder me' (p. 30). He has turned himself into a passive object, a boat left in charge of an active but incompetent helmsman. Later in the book, at the end of three months' abstention from alcohol, Brummy Tom persuades him to go to the pub. He excuses his backsliding by introducing an extraneous, sociological reason for it: ' "Brum," I said, rather bitterly, "a tee-totaller who lives in a common lodging house is to be heartily despised, for he shows himself to be satisfied with his conditions" ' (p. 190). The implication is that drink amongst his fellow tramps is a deliberate act of protest against their oppressed status. But the book has long since shown how much a matter of choice Davies's 'oppression' is. So he uses a different wholly diversionary tactic to justify his subsequent reversion to what he euphuistically calls 'a number of long sleeping draughts': 'The following morning some of the lodgers were telling of murder cries heard just after midnight, but I praised the power of Bacchus that I had not heard them' (pp. 190–1). The weakness of this piece of rationalization does not belong only to his earlier self. By repeating it without comment he is giving it implicit condonation in the present.

Style and language, then, are instruments in a strategy of evasion that is central to the make-up of this basically shy man. Ultimately Davies has no desire to uncover that naked self whom he dresses in such seemly language. His conception of autobiography is that of a good adventure story that happens to be true. As we have seen, any narrative reliant on memory cannot avoid some untruths. But in addition Davies is not interested in exploring truths of meaning. He is not sufficiently separated from his earlier self to allow the unimpeded interplay between past and present selves which lies at the heart of good autobiography. This is frequently the case with autobiographies written early in an individual's life. Interplay of that kind offers a more complex and less incomplete form of truth than does an autobiography which assumes a unity of identity between protagonist and narrator for the sake of concentrating attention on the plot line. Davies's *Autobiography* is not so much untruthful as truthful only in a confined sphere – that of factuality. But facts in sequence call out for interpretation. The nature of the genre repeatedly confronts Davies with questions about his constant reversion to heavy drinking or his dislike of paid work. He evades the questions by clothing himself in the guise of a natural with a natural's artless speech. Yet his supposedly natural passages of self-recrimination, like the supposedly natural language he employs, are anything but artless. By trying to confine the truth about his life to the facts of his past history he has been forced to fudge over or avoid those other truths concerning not 'what?' but 'why?' which clamour for attention the more they are repressed. If

The Autobiography of a Super-Tramp is typical then a concentration on factual truth in autobiography appears necessarily to produce a particularly glaring simplification of that chimera, the whole truth of a life, which obviously can never be told.

At least Davies's *Autobiography* has helped to establish the parameters of the 'factual' approach. His adoption of this approach has not prevented him from writing a bestseller, a form of picaresque autobiography which has a similar appeal to that of adventure books. At the same time the genre has shown itself to be strongly resistant to the simplifications produced by assuming a purely factual approach to one's life history. A similar approach has been adopted and more fully developed by one of the century's most rigorously honest writers, George Orwell. His exceptional determination to rid himself of his inherited prejudices in order to get to know the underside of life, not just around him but within him, leads him to confront more directly than Davies was prepared to do the generic pressure to attach meaning to fact, to face, that is, the inescapable way in which an accumulation of facts creates willy-nilly a crude meaning of its own.

George Orwell: *Down and Out in Paris and London*

Down and Out in Paris and London (1933), like *The Autobiography of a Super-Tramp*, was its author's first published prose work. Like Davies's, it was autobiographical and recounted Orwell's immersion in the company of society's rejects, including tramps. So it provides a particularly illuminating comparison to Davies's *Autobiography*. It also constitutes the nearest approach Orwell came to writing straight autobiography, excluding his essay-length reminiscence of his experiences at his preparatory school, 'Such, Such were the Joys'. Unlike Orwell's later two factually based books, *Down and Out* covers at least two years of his life and concentrates for the greater part on his personal experiences. By comparison *The Road to Wigan Pier* (1937) and *Homage to Catalonia* (1938) cover only a matter of months and come closer to documentary journalism than does the more personal *Down and Out* which mainly confines its sociological observations to four chapters at the end of its two sections.

W. H. Davies was not alone in reacting to Orwell's book as 'all true to life, from beginning to end.' 'This is the kind of book I like to read,' Davies wrote in his review for the *New Statesman*, 'where I get the truth in chapters of real life.'[17] But how real? The first twenty-three chapters of *Down and Out* concentrate on Orwell's final ten weeks in Paris between October and mid-December 1929. In October he ran out of money and, after several days with little or no food, he went to work as a dishwasher, first at the Hôtel Lotti in rue de Rivoli, then at the Auberge. The last fifteen chapters concertina Orwell's self-imposed wanderings as a tramp in England during the winter of 1927–8 and the spring and summer of 1930 into a period of four weeks immediately following his return from Paris. In his Introduction to the French edition of *Down and Out* he openly confesses that he has rearranged the facts. At the same time he offers as his excuse the fact that any factual account has to be a selective arrangement by the mere act of narration, something of which Davies appears less aware. As Orwell explains it:

As for the truth of my story, I think I can say that I have exaggerated nothing except in so far as all writers exaggerate by selecting. I did not feel I had to describe events in the exact order in which they happened, but everything I have described did take place at one time or another.[18]

Two years later Orwell was to qualify 'everything' by 'nearly'[19] in *The Road to Wigan Pier*. But the point Orwell has grasped is that facts acquire meaning by the way in which he retrospectively chooses to order them. That meaning therefore pertains to the author at the time of writing and usually offers a different perspective on the facts to how they were experienced at the time.

Biographical research has shown that Orwell's final sentence does require qualifying in several respects. Thus one friend of his has claimed that it was not an Italian compositor as in *Down and Out* but a Parisian girl he had picked up who robbed him of virtually all his money and forced him to seek work in a hotel.[20] Then no one can trace a friend resembling 'B' in the book where he pays for Orwell's return ticket from Paris and arranges a tutoring job for him after he has got back. It is far more likely to have been his parents who bailed him out, especially as the tutoring he did on his return home from Paris to his parents' house in Southwold was just across the river from them by ferry. Then, unlike the order in which events are described in the book, he began his tramping before his stay in Paris where he wrote a number of articles about them including 'The Spike' the contents of which he redistributed between Chapters 27 and 35 of *Down and Out*. His tramping was the result of a conscious choice on his part to discover how 'the other half' lived, not the result of exigency as he makes out in the book. Besides his parents he also suppressed the supportive presence of his Aunt Nellie and her Esperanto-speaking lover in Paris. He seems to have begun life in Paris as their lodger. After moving out to the cheap hotel in which he is discovered in the opening pages of *Down and Out* he continued to see his aunt on and off throughout his Parisian stay. This means that at any time he could have appealed to her for help and that the real poverty and near starvation he suffered in the final few months of 1929 were avoidable had he chosen to ask her for help. On the one hand he seems to think it essential for a creative artist to experience poverty in order to write about it. On the other hand he does not want to appear to recognize that his experience of poverty was crucially different in kind from those who had no option. He may have persuaded himself that his sense of pride ruled out the possibility of asking his family for help. But the omission of all mention of the option creates an over-simplified picture of the complexities of the actual situation he found himself in.

Alternatively it could be argued that he chose to omit these facts for artistic reasons. 'Poverty is what I am writing about,' he announces at the end of the first chapter.[21] He may well have thought that the question of whether the poverty was self-imposed or involuntary would only distract attention from the theme of the book. This artistic consideration may also have been why he virtually expunged from the book mention of the fact that he was living in Paris to establish himself as artist, a creative writer. He does admit to this in the Introduction to the French edition. But in his first published book he shows an understandable reluctance to advertise his early literary failures. So literary-cum-social considerations can conflict with and even override those of purely 'factual' autobiography.

Within these self-imposed limits Orwell's account of his immersion in various forms of poverty is strikingly fresh and original. The book is littered with personalized facts – how much money he has at any time, what hours he worked, how many miles he covered per day, and the like. His first experience of being broke and hungry, like his first appearance in tramp's clothes, entails unflattering revelations that have the effect of enhancing the authenticity of the narrative. The lies poverty reduces him to telling, expensive lies at that, the squalid subterfuges and minor economic disasters, the boredom and apathy are all told from first-hand experience. Orwell admits, for example, pouring his last 80 centimes' worth of milk away after flicking a bug into it, a self-defeating and effete reaction considering his need at the time. He shows the same middle-class priggishness when confronted at his first English lodging house with wash basins streaked with grime, preferring to leave unwashed. Or one remembers the shock with which Orwell, after changing into a tramp's clothes, realized that the beggar he saw approaching him in the reflection of a shop window was himself. Here he briefly touches on the reflective nature of the autobiographical act itself. Eric Blair observes George Orwell writing about a younger Eric Blair observing his own transformation from one class to another, from the socially acceptable to the socially unacceptable, from the product of English public-school education and middle-class conditioning to the representative of all he feared in himself and human nature.

As an observer of a world alien to his readers Orwell proves to be a credible reporter. There is no attempt to turn the down-and-outs he meets into noble victims of a callous social system. Boris, the Russian *émigré* he sees most of in Paris, harbours a White Russian officer's ugly anti-Semitism. Jules, the Magyar second waiter at the Auberge, refuses to work on principles derived from a Communist's contempt for all bourgeois bosses. Paddy, the Irish tramp he spends most time with in the English section, is full of self-pity, has 'a low worm-like envy of anyone who was better off', is zenophobic, 'abject, envious, a jackal's character' (pp. 152–3). 'Nevertheless,' Orwell continues, 'he was a good fellow, generous by nature and capable of sharing his last crust with a friend' (p. 153). The conclusion that Orwell reaches reflects his as yet unpoliticized social and political outlook at this time: 'It was malnutrition and not any native vice that had destroyed his manhood' (p. 153). In passages like these the distance that separated Orwell the observer from his fellow down-and-outs nicely parallels the distance that separates Orwell the narrator from Orwell the protagonist of his autobiography.

This sense of distance is responsible, too, for Orwell's discovery that class distinctions persist just as tenaciously among the working-class employees of the Parisian 'Hôtel X' (Hôtel Lotti) as among its rich customers. He makes much of the double door separating the dirty little scullery in which sweating waiters slithered about on layers of discarded food and the spotless dining room where customers sat in their splendour. He spends longer on what he calls 'the elaborate caste system existing in a hotel' (p. 70), from the high-caste waiters to the untouchables, the *plongeurs*, to which he belonged. What emerges from his description of the two worlds is an intellectual contempt for the average customer who is 'put to shame by having such an aristocrat to serve him' (p. 69), but a deep physical revulsion for the 'disgusting filth' of the scullery and the 'sweaty armpits' of the waiters. It seems that, for all his slumming it, he could not escape the puritanical desire for cleanliness that he inherited from his childhood. Orwell's scrupulous attempt at remaining objective does give him original insights into the motivation of each class of worker, insights which are not to be found in Davies's book. One such insight is his realization that a waiter's work 'gives him the mentality, not of a workman, but of a snob' (p. 76). Equally unexpected is his conclusion that the 'constant war between different departments also made for efficiency, for everyone clung to his own privileges and tried to stop others idling and pilfering' (p. 79).

Observations like these suggest that Orwell saw himself as some kind of applied scientist or anthropologist whose task was to efface himself and his opinions in order to report back to his own social class the mores of the underdogs of life. Where the pressure of the genre compels him to focus on himself he attempts to view himself with the same detachment, like the reflected image of himself in tramp's clothing that momentarily presented itself as that of a complete stranger.

But when the self being observed overlaps with the self making the observation the objective stance is untenable. Perhaps the most obvious instance of this occurs in the chapter in which he suspends the action to make a series of sociological observations about his Paris experience. When he tries to get beneath the economic reasons for perpetuating useless work (as he sees that of the *plongeurs*) he offers a psychological explanation that boils down to rich people's innate fear of the mob: 'The mob (the thought runs) are such low animals that they would be dangerous if they had leisure; it is safer to keep them too busy to think' (p. 119). This 'superstitious fear', he argues, is based on the mistaken belief that rich and poor are differentiated by anything other than their incomes. This is more a form of wish–fulfilment on Orwell's part than an impersonal judgement. Orwell desperately wants to believe that the conditioning of class and education can be eradicated by, in his case, immersing himself in the world of the poor. Yet his autobiographical account of his voluntary experience of poverty repeatedly illustrates how impossible it is for him to become indifferent to dirt and body odour. No matter how hard up he finds himself he considers it necessary to waste money in a dozen different ways 'to keep up appearances' (p. 17). With his background, to find a bug in his milk means doing without the milk. His experience of poverty is different from that of his fellow paupers in a whole variety of similar circumstances. His sociological thesis runs counter to his personal history. The genre he employs has the effect of highlighting these contradictions. His personal investment in finding no difference but income between rich and poor tends to invalidate all his general conclusions and even to cast doubt on the objectivity of his thought-provoking observations on, for instance, the snobbery of waiters as a class. Doesn't it suit his thesis too neatly to show them indulging in the same human foibles as their rich customers? Isn't the autobiography, in fact, the product of an internal argument which Orwell badly wants to go in the particular direction of his thesis? Yet his own personal experiences, to which the genre constantly compels him to give truthful expression, run in the contrary direction. The autobiography, then, reveals an elementary divide within its author which he is powerless to conceal.

The desire to find evidence for his thesis among the personal observations he accumulated during his periods with the down-and-outs has led him to adopt an attitude towards his depiction of character that owes more to fiction than to the tradition of autobiography. In his Introduction to the French edition of the book he admits to a deliberate bias in his portraits:

I have refrained, as far as possible, from drawing individual portraits of particular people. All the characters I have described in both parts of the book are intended more as representative types of the Parisian or Londoner of the class to which they belong than as individuals.[22]

In making typicality the principle of his selection he is simultaneously allowing himself to dwell on those aspects of the rich and the poor which illustrate his argument that they are equal in everything but income. Yet his sociological contention rests on a particularly non-realistic device borrowed from the dramatic farce, that of caricature. 'Boris' is evidently one such representative type. Although in *Down and Out* Orwell claims that he met Boris in the Hôpital where he himself was recovering from pneumonia, when later he wrote a documentary account of his stay at the Hôpital, 'How the Poor Die', the only patient who, like Boris, has a game leg is a little red-haired cobbler. Boris then may or may not have had a principal sitter, but there are times when he comes close to caricature. His plans to turn criminal, his ex-mistress's letter, his diatribe against the Jew who rents his bug-ridden bed, all have the taint of that melodramatic streak in the early Orwell that reveals itself as pure fantasy at other points of the book. Two such points are his highly coloured portrait of Charlie with his tales of his night's orgiastic despoliation of a young virgin, and of Roucolle, the rich miser cheated of 6,000 francs which he paid for face powder thinking it to be cocaine. Admittedly these are recounted by characters in the book and not by Orwell in person. But their garish quality, compared to the reportage of the bulk of the book, undermines the authenticity of Charlie, the character who tells them. It is as if the literary self that Orwell largely concealed from the reader is finding surreptitious expression through the mouth of Charlie.

This technique of seeking out the typical from a number of individuals and then imbuing it in one representative figure works more convincingly in the case of Paddy. 'Paddy' is an amalgam of Nobby in 'The Spike' and the Irish tramp with whom he wandered for several days (not two weeks as in the autobiography) through the northern outskirts of London mentioned in *The Road to Wigan Pier*. Paddy appears convincing partly because he is set among other individualized characters, like Bozo, the pavement artist who dismisses Paddy as 'a tea-swilling old moocher, only fit to scrounge for fag-ends' (p. 165), or the unemployed young carpenter who, despite being on the road for six months, dissociated himself from his fellow tramps whom he repeatedly calls 'scum'. Another reason why Paddy appears more plausible than Boris is because Orwell is more successful in reproducing his Irish brogue than the English equivalent for Boris's Russian pronunciation of French. Caricature is most often to be found at the comic extreme of the dramatic spectrum, in farce. Yet Orwell's literary employment of language appropriate to his caricatured portraits at times unexpectedly adds rather than detracts from the authenticity of his narrative.

In this first published book of his Orwell nevertheless still vacillates between the transparent prose style he aspired to and the purple passages and 'humbug' that he was fighting to shed. Many reviewers of *Down and Out* praise Orwell for his 'quiet, level tone', his 'vocabulary suited to his subject'.[23] However in his late essay of 1946, 'Why I Write', he claimed that he achieved his ideal of transparency (prose in which one manages 'to efface one's personality') whenever he was motivated by 'a *political* purpose'. If one interprets 'political' in its widest sense, then one can draw a distinction between most of the impersonal descriptions of life in Paris kitchens and English spikes, and his purple descriptions of Charlie's sadistic night with a young virgin or the melodramatic story of Roucolle's come-uppance. Occasionally, however, the garish red ambience of the room in which Charlie despoils his virgin spills over into lurid descriptions of the rue du Coq d'Or or of the kitchen in the Hôtel X, 'a stifling, low-ceilinged inferno of a cellar, red-lit from the fires' where cooks swore and sweated over furnaces stoked by scullions naked to the waist (p. 57). As Stephen Greenblatt has pointed out, Orwell's non-realist prose style here is employed to construct 'an image of modern hell and a sweeping condemnation of capitalism'.[24] Orwell, in fact, is led to use purple prose because he is concerned to make a political point, quite the reverse of his claim in 'Why I Write'.

There is a basic contradiction between the desire 'to efface one's personality' for political motives and the desire to report back on one's *personal* immersion in the society of down-and-outs. Purple passages, representative characterization and chronological rear-rangement of material together amount to a concerted assault on 'factual' realism. Nowhere is this contradiction more apparent than in Orwell's ambiguous attitude towards his portrayal of himself. Whatever the motives, the suppression of the support he received from his aunt in Paris and his parents back in England creates the false impression that he was driven by sheer indigence to work as a washer-up in Paris, and to tramp the roads of south-east England. To have confessed that in reality he exchanged clothes at the houses of various affluent friends before and after his voluntary tramping expeditions would undoubtedly have diverted readers' and critics' attention from the insights he had acquired on the road.

In using an autobiographical mode to focus on a set of little-known facts, Orwell was, without realizing it, creating an internal conflict between the genre's natural concentration on personally experienced truths and his 'political' desire to present without bias a picture of life among the poor and the down and out. Yet this picture of society's victims and rejects acquires authenticity only when it is seen to be painted at first hand by a fellow sufferer. We need to believe in the narrator-cum-protagonist, to share his feelings of fear, disgust, outrage, in order willingly to accept his findings. Autobiographical reticence paradoxically undermines narrative credibility. This reticence of his is the result of a number of factors: his desire not to offend his parents; his wish to concentrate on the theme of the book, poverty; above all, a resistance to the genre's drive towards self-revelation. Why else should he attribute to some fictional German *plongeur* his own exploit (as Bernard Crick has shown) in which he stole peaches asked for by a customer of the hotel rather than return empty-handed?[25] Is it also the same resistance that makes him admit only indirectly to being an old Etonian when he is accosted by a drunken fellow Etonian in a cheap lodging house? His aspirations towards impersonality within a personally oriented narrative suggest the presence of an unresolved, because as yet unrecognized, personal investment in the material he is handling.

Yet where Davies throughout his writing life never came to terms with his own buried motives for tramping and begging, Orwell did face his hidden need to immerse himself in low life in the autobiographical Chapter 9 of *The Road to Wigan Pier* written four years later. There he admits to the immense burden of class guilt which had accumulated in him during his five years in the Indian police. In violent reaction, he wanted 'to get right down among the oppressed, to be one of them and on their side against the tyrants'. Hence his neurotic need to join 'the lowest of the low', thereby 'getting out of the respectable world altogether'. In the same chapter he admits that his account in *Down and Out* of exchanging his clothes for those of a tramp at a rag shop was not strictly accurate. In fact he bought 'the right kind of clothes and dirtied them in appropriate places'. Subsequently he changed into them at a friend's house. Significantly he describes his first entry into a common lodging house as a 'baptism' and a 'release'.

Later still Orwell was to trace the origins of his neurotic sense of guilt back to his childhood at St Cyprian's Preparatory School in 'Such, Such were the Joys'. It seems improbable that he would have achieved greater clarity of style and integrity in his subsequent writing had he not pursued his motives for those years of slumming in Paris and London back to these childhood experiences. His capacity for objective reportage increased in direct proportion to his capacity for subjective self-discovery. It is his attempt to minimize his role as active participant in *Down and Out* that is responsible for the presence of distortion and bias in this purportedly objective narrative. His unrecognized subjective needs leave him insufficient space for a true perception of otherness, of the factual world beyond the self. 'Factuality', however limited, is dependent on the self-awareness of the autobiographer, because autobiography is a relative act in which the observer/participant necessarily gives meaning to the nature of all he observes and engages. The only effective counter to this unavoidable subjective bias is for the autobiographer to become aware of the bias and to share this awareness with the reader. In this way not only has the autobiographer more chance of correcting the subjective slant of his observations of the world outside himself, but also the reader is better placed further to correct the unperceived distortions of the writer. This, I would argue, is why the genre reaches its full potential only when it overtly embraces the subjective element implicit in all autobiography.

2

Internal Verification

The truths which can be revealed in autobiography, it has been shown, are broader and more complex than sheer factual veracity. But they are still encompassed by the expectation that an autobiographer will do his best to adhere to the facts as he knows them. At once one finds oneself in the minefield of memory. Memory is normally the principal source of raw material for the autobiographer. Even when he has had the help of letters or diaries, it is still memory that gives life to the incidents selected for their significance from such documents. What imbues them with significance is likely to be the strength and vividness of the memories they evoke. So that memory is an unconscious agent of selection. And memory is notoriously unreliable.

Most good autobiographers are all too aware of the fallibility of this faculty on which they are forced to rely so heavily. It is one of the features of the genre, built into its conception, what Roy Pascal calls 'a necessary condition of it'.[1] Some writers readily abandon themselves to its power, confident in the assumption that 'memory is a great artist',[2] as André Maurois puts it. Others feel more ambiguous about it. Near the beginning of his autobiography G. K. Chesterton spends several pages demonstrating to his readers the paradoxical nature of this faculty on which he is necessarily dependent:

Really, the things we remember are the things we forget. I mean that when a memory comes back sharply and suddenly, piercing the protection of oblivion, it appears for an instant exactly as it really was. If we think of it often, while its essentials doubtless remain true, it becomes more and more our own memory of the thing rather than the thing remembered ... This is the real difficulty about remembering anything; that we have remembered too much – for we have remembered it too often.[3]

This need to educate his reader in the complexities of the genre is not unusual. Yet even Chesterton asserts without hesitation that on its first occasion memory 'appears for an instant exactly as it really was'. Does it? At least one school of thought asserts that the unconscious not only buries memories but, 'literally, ruins the objects that come to inhabit it'.[4] According to this argument any act of memory is therefore an act of archaeological reconstruction. The object of such reconstruction is inevitably other than the original event or feeling. Yet it is all that is available to the self seeking its own elusive past. Autobiographers who subscribe to this view are consequently forced to accept and indeed exploit the creative, even inventive nature of the autobiographical act. They make of their lives artefacts and ask the reader to understand them as such.

Freud regarded conscious memory as that part of one's past still available to consciousness after the ego has repressed those memories likely to produce anxiety. Seen from this point of view memory is nothing more than 'a pander to one's ego',[5] as A. E. Coppard observed in his autobiography. Since many autobiographers see it as their task to portray the ego, or conversely see it as an impossible task to portray the unconscious self, this bias in the memory has not prevented them from relying predominantly on it for their raw and finished material. There are, however, ways of countering its egotistical bias and autobiographers have resorted to one or more of these strategies. A number, for instance, make use of personal and historical documents of various kinds – diaries, letters, accounts, newspapers, books and the like – to overcome the distortions of rerememembering. Others seek indirect access to the uncontaminated truths of the unconscious by retelling their dreams either accompanied by interpretations or not. Yet others allow the act of writing itself to uncover by association memories left buried up to the moment of writing. However, for the most part autobiographers are resigned to the impossibility of recovering their past life as it was experienced in the past. They are ready enough to use memory to re-create their past in the present, on the grounds that they are being true to their *present* conception of their selves rather than to their irrecoverable past. Henry Green is representative of numerous autobiographers who choose to rely on a memory they know to be unreliable because it nevertheless provides valuable evidence of their later self-image. 'Most people,' he writes, 'remember very little of when they were small and what small part of this time there is that stays is coloured ... and retouched enough to make it an unreliable account of what used to be. But while this presentation is inaccurate ... it does set out ... what one thinks has gone to make one up.'[6] This leads him further to insist that any external aid to memory such as being told what happened to one as a child 'has no bearing on what one has experienced', however accurate it might be.

Does this modern tendency in autobiographers to flaunt the subjective experience of the present in the face of potentially objective accounts of the past break the age-old pact between readers and writers of autobiography? Has all claim to any mutually understood form of truthfulness been imperceptibly jettisoned in the course of this century? I think not. This is because the great majority of autobiographical writers take for granted the presence of certain internal criteria by means of which the reader is able to assess how honest, how 'truthful' they are being. By 'internal' I mean internal to the text, intra-textual. These criteria vary from intentional warnings to the reader – not, for example, to expect the whole truth – to the sublimal, give-away clues that point to the way in which the writer is censoring, distorting or sometimes inventing material to enhance his self-image.

Typical of the more conscious strategy is A. E. Coppard's admission on the first page of his autobiography: 'There have been episodes in my life, important and privately fascinating occasions, which not even the prospect of an eternity of hellfire would induce me to reveal.' This is true of all autobiographical writers, whether they admit it or not. Coppard continues: 'There are other things which I may endow – I know it, I shall – with a spurious glister.'[7] That too is a near-universal failing among autobiographers. At a less obvious level there is the motivation of the writer. When, for instance, Wyndham Lewis announces that his main aim in writing *Rude Assignment* ('A narrative of my career up-to-date') is 'to spoil the sport of the irresponsible detractor, to improve my chances of some day not being too much lied about',[8] the reader is instantly alerted to a polemical, self-justificatory bias in the book, a bias of which he duly finds evidence in abundance.

But there exist in all autobiographies less overt clues to the degree of truthfulness practised by the writer. In the first place the individual responsible for telling the truth is simultaneously describing himself, thereby providing us with a means by which to gauge what degree of honesty we can expect from such a personality. In his pioneering study of the origins of autobiography in antiquity Georg Misch points out how impossible it is for any autobiographer to conceal his true nature from the reader: 'Even the cleverest liar, in his fabricated or embroidered stories of himself, will be unable to deceive us as to his character. He will reveal it through the spirit of his lies.'[9] Then there is the question of the rigour with which a writer pursues his quest of the self, his degree of self-awareness and self-criticism. In most cases, Roy Pascal argues, the untruth of autobiographies 'is evident . . . in their lack of probing, in their relaxed mood'.[10] When, for example, Geoffrey Grigson asserts in his autobiography that 'childhood, in retrospect, is a tedious affair, which goes much to pattern',[11] one is at once alerted to an air of self-deception. To see childhood as conforming to a pattern is to deny or forget the intensity of an individual's self-awakening which is a unique experience for each child. This leads me to suspect that Grigson has adopted a stance of cultivated perversity which is reinforced by his assumed ennui ('a tedious affair'), hardly natural for a man at the height of his powers in his mid-forties. Style, language and form all offer the reader additional means of assessing to what extent the writer is living up to his standard of truthfulness. As J. N. Morris suggests, the autobiographer's 'manner has to do not only with how he tells his truth; it is itself the truth'.[12]

The way that these internal criteria operate within an autobiographical text becomes apparent in an autobiography that is at the same time probably its author's highest claim to fame and notorious for its unreliability – Frank Harris's *My Life and Loves*.

Frank Harris: *My Life and Loves*

Harris began writing the first volume of *My Life and Loves* in 1921 when he was 66. He continued intermittently to work on his autobiography until he died ten years later with the fifth volume still incomplete. Volume I was published privately in 1922, Volume

II in 1925, Volumes III and IV, originally one long volume, appeared in 1927, while the episodic Volume V was published posthumously in 1958 edited and enlarged by Alexander Trocchi. A definitive edition[13] of all five volumes in one book edited with notes by John F. Gallagher was first published in the States in 1963. This runs to almost 1,100 pages and spans Harris's life from 1855 to 1928. I will concentrate on the first two volumes covering his life to the late 1880s. These constitute the better half of the autobiography since they have a stronger narrative line and are less subject to digressions than the last three volumes.

What an astonishing career this book purports to unfold. Born in Ireland, sent to boarding school in England, Harris ran away to the United States at the age of 16. According to his account, he worked in New York as a bootblack and underwater worker on the foundations of the Brooklyn Bridge, as a night clerk, bookkeeper and eventually steward at a hotel in Chicago, left to become a cowboy running cattle from Texas to Kansas City and Chicago, and ended up in Lawrence, Kansas, attending courses at the University, making a fortune letting billboards, and gaining admission to the bar. At 18 he left for a spell at the Sorbonne in Paris, then taught at Brighton College, summered in Russia, and went on to spend about four years studying in different German universities. Back in London he became editor of the *Evening News* at the age of 28. Quite apart from his reported sexual feats during this time, he has packed more experience into these early years of adult life than most people manage in a lifetime and recounted it with immense gusto. Read as one man's experience of Western civilization during a period of critical change, Harris's autobiography spans an immense range with great skill.

Volume I was written in the hope of restoring his depleted fortunes. At the same time the autobiography was extremely ambitious in its conception. 'I want to . . . write the greatest Memoirs that have ever been written,'[14] Harris wrote to Mencken after his first volume had been greeted with near universal outrage and abuse. In his Foreword to Volume III he unashamedly compares his unfinished autobiography to the works of Anatole France, Verlaine and James Joyce (p. 520). Writing to Hesketh Pearson, he claimed that the chapter describing his seduction of Laura, his first real love, 'reaches in the poetry in the end an even more intense expression of passion than Shakespeare has reached in Othello'[15] — a highly appropriate parallel in the circumstances. His two principal models in the genre were Rousseau's *Confessions* and Casanova's *Memoirs*, although he claimed that his book was 'more honest' than either. Harris persuaded himself that as an autobiographer he was a second Moses, using his life as the text for his vision of the future: 'What I preach today amid the scorn and hatred of men will be universally accepted tomorrow; for in my vision, too, a thousand years are as one day' (p. 7). Drawing on the more sexually outspoken language of Chaucer and Shakespeare, he saw his mission in writing his autobiography as reversing the trend of modern civilization towards a disastrous Pauline puritanism and as restoring wholeness by combining pagan with true Christian ideals. *My Life and Loves*, he claims in the Foreword to Volume I, 'is the first book ever written to glorify the body and its passionate desires and the soul as well and its sacred, climbing sympathies'. He is not just a successor in his own eyes to Moses and Jesus; he is also his own evangelist. 'I have tried to write the book I have always wanted to read,' he concludes, 'the first chapter in the Bible of Humanity' (p. 9).

These are high claims indeed. Yet the way in which he expresses his ambitions leaves an indelible impression of vanity. In the opening pages he readily admits to what he calls 'my besetting sin', 'the vanity in me which had already become inordinate and in the future was destined to shape my life and direct my purposes' (p. 16). Ambition and vanity coexist in Harris's autobiography, both alerting the reader to the possibility of distortion and falsification. Both qualities are responsible for his tendency to claim parity with the great. On more than one occasion he claims to have anticipated ideas which the public mistakenly suppose were originated by more famous contemporaries of his. His schoolboy interpretation of Shakespeare's Shylock, he claims, included 'the very piece of "business" that made Henry Irving's Shylock fifteen years later "ever memorable" according to the papers' (p. 49). Arguing about wages theory with his fellow cowpunchers over the camp fire, Harris, as he professes to have realized ten years later when reading Herbert Spencer, 'had divined the best of his sociology and added to it materially' (p. 89). Similarly he claims to have arrived at the same conclusion as Goethe did concerning the rival demands of

individualism and socialism 'years before reading the fragment of
the play *Prometheus*, that contains the deepest piece of practical
insight to be found anywhere' (p. 251). In all these instances Harris,
the neglected genius, is in effect rewriting history in order to insert
retrospectively his own supposed contributions to Western
thought. Clearly he badly wanted to be thought of as an original
and profound thinker rather than as the ebullient literary entrep-
reneur he had become. It is a suspect process in itself which
becomes doubly suspect when one takes into account his over-
wheening self-esteem. To find him denying at the close of Volume II
('I have no pride whatever . . .') what he confessed to at the
beginning of Volume I only serves to increase one's distrust of these
assertions in which ambition and pride have so obvious an
undeclared interest.

Similar doubts arise from his almost paranoiac defensiveness in
the Forewords to Volumes I and II. In Volume I he brazenly admits
that the frankness with which he has treated the sexual content is
due to his urgent need to make money: 'If America had not
reduced me to penury I should probably not have written this book
as boldly as the ideal demanded' (pp. 1–2). The ideal (to be 'a brave
soldier in the Liberation War of Humanity', liberation, that is, from
sexual inhibition) is rather diminished by the money motive and
further clouded by Harris's conviction that his lack of money is due
to an ignorant public on whom, he writes, 'I turn at length to bay'
(p. 2). The analogy to a trapped animal about to fight its
persecutors to the death betrays a hysterical attitude to the entire
autobiographical undertaking that is unlikely to be conducive to
accurate reportage. The fact that he identifies truthfulness especially with sexual explicitness immediately arouses one's suspicions
about the accuracy of the sexual episodes in the book. If it is to sell
well these episodes need to be exceptional and told with frankness.
If they weren't exceptional in life then they will need to be
heightened. One's suspicions are further aroused when Harris tries
to conceal in the Foreword his already admitted pecuniary motives
behind a smokescreen of selfless benevolence. He wants to warn the
young and impressionable, he writes loftily, about the dangers of
sexual misuse so that they might preserve their potency into old age
(pp. 3–4). His main desire, he claims, is 'in this way to increase the
sum of happiness in the world' (p. 4). In the body of the book the

spasmodic didactic passages he inserts for the sake of the young sit as uncomfortably as do Davies's exclamations against drink and indigence.

Far from making him a quick fortune the first volume was a commercial disaster, confiscated by authorities throughout Europe and the States. Harris's persecution complex was reinforced. He was 'boycotted and disgraced, his books seized and burned' (p. 219). He had, his critics claimed, committed 'literary and moral suicide' (p. 220). Just as Christian martyrs were prepared to die for the sake of their souls, so Harris implies that he is ready to lose all for the sake of preserving a frank account of the self. The degree to which he needs to see his life in terms of heroic drama is equally in evidence in his portrayal of the women in his life. His claims on their behalf smack of the world of romance and legend where wish–fulfilment is indistinguishable from reality: 'The perfume of [women's] flesh is sweeter than all the scents of Araby, and they are gracious-rich in giving as crowned queens' (p. 222). The rich language of this make-believe world has drawn him into its spell. Yet in the Foreword to Volume II he is forced to come to terms with the real world around him. There he admits that, in response to the near-universal vilification he received after the publication of the sexually explicit Volume I, he will 'not be as contemptuous of convention' nor as 'exact and painfully true' in the next volume as he was in the first (pp. 225, 223). His 'Holy Spirit of Truth' has fallen victim to the exigencies of the marketplace. Volume II shows a marked toning down of his sexual vocabulary, while the virtues of chastity are suddenly extolled.

Robert Pearsall has calculated that out of the all but 1,100 pages of the American edition of *My Life and Loves* less than 65 pages are devoted to the erotic incidents with which the public associates the book. Nevertheless Harris, judging from his forewords, obviously considered the sexual element a vital ingredient. His attitude towards sex is a peculiar blend of modern liberation and Victorian ignorance. He firmly believed that too much sex can be harmful. Excessive masturbation reduced one schoolfellow to tears and nervous shakes, he reports. Wet dreams, he reckons, had a disastrous effect on his own performance at the high jump and are held responsible for the death of his friend and mentor, Professor Byron Smith. His belief in the use of a syringe to prevent conception is equally archaic. But these features, however ridiculous, at least ring true. What soon arouses the reader's suspicion are Harris's Olympic feats of sexual gymnastics, because these reflect to the greater glory of our questing hero. Before one typical bout with one of his several mistresses in Lawrence, Kansas, she tells him: 'You and you alone satisfy the insatiable; you leave me bathed in bliss, sighing with satisfaction, happy as the Queen in Heaven!' (p. 139). Besides using his female mouthpiece to praise his sexual expertise, Harris allows the rhetoric to remove the situation from the real to the realms of medieval alliterative verse. Within a page his skilled love-making brings about her virtual breakdown from an excess of pleasure. So it comes as quite a shock to find that by the end of Volume II he seems to have forgotten the existence of this married mistress when he boldly asserts: 'I never coveted another man's wife' (p. 508). His change of theme from one of sexual explicitness to one in praise of chastity produces a complete reversal of the facts. Facts are frequently dispensed with in this way when they come into conflict with Harris's urgent desire to win approval from his public.

Clearly Harris attaches more importance to the thematic interpretation of his material, even if these interpretations happen to be in conflict with one another, than to the material in its own right. His current state of mind invariably supersedes his historical sense of the past. Internal contradictions between Volumes I and II are correspondingly frequent. For instance the same married woman in Lawrence, Kansas, introduces him to oral sex. Yet subsequently in Volume II he beds in Athens a Mme M— who out of fear of conceiving insists on oral sex. After that, Harris writes, 'We always practised the same game she had been the first to teach me' (p. 276). Then there is the first occasion on which he awoke to the beauties of nature. In Volume I this is after the 14-year-old Harris has been touching up Lucille, the French governess at the local vicarage. In Volume II, on the other hand, it is when, at the age of 13, he is innocently lunching at Lady Wynn's. Volume II, he tells us in the Foreword to it, is inspired by the practice 'of total abstinence from all sex-pleasure for years' (p. 224). By Chapter III he is bedding an English girl at Frankfurt and only fails to consummate the affair for fear of hurting her. On finding that she only despises him for this act of self-control he resolves 'not to be such an unselfish fool next time' (p. 247). In Chapter VI comes the episode with Mme M—. In the next chapter he fails to seduce his Irish landlord's daughter, but not for want of trying. Then follows his first real love affair with Laura. But when he catches her in another man's company he rushes home and promptly beds his maid. Finally there is the rich cocotte he picks up in Paris but eventually feels compelled to give up because she is too weakening. That is hardly 'total abstinence'. His wish to placate his irate readers is not strong enough to repress his greater need to win approval for and from himself. So, by implication, how reliable are the accounts of his sexual liaisons themselves? Can one really take seriously his claim in Volume V that 'there is no selective heightening of sensation nor any attempt to exaggerate sexual pleasure in any description' (p. 849)? When one remembers that he has already confessed to a 'loss of virility' as a result of which 'the glamour has gone out of life' (p. 507), one cannot help speculating whether the rewriting of his life in heightened retrospect did not help restore the glamour that gave validation to his life. His reputation in eclipse, the old man of letters seized on autobiography as a way of staging a comeback to his former glory as hero of his own psychodrama.

Harris himself provides evidence which confirms this suspicion that his memory functions in a highly selective and inaccurate way. In his Foreword to Volume I he wishes he had begun his autobiography five years earlier before he had 'been half-drowned in the brackish flood of old age and become conscious of failing memory' (p. 9). On the next page he none the less declares: 'I take memory in the main as my guide.' He further proceeds to boast of his prodigious feats of memory as a child, feats which are so staggering that the average reader must view them with considerable scepticism. He claims to have learnt by heart *Paradise Lost* in a week and *The Merchant of Venice* in a day, each by a single careful reading. Later, in his early twenties, he reckons to have mastered French so that he could 'understand everything said by the end of the first week' (p. 194). Yet five years later, back in France, even after working hard at French he 'did not attempt to master it' (p. 288). His claim to understand and be understood in Italian in a fortnight is undermined by his scarcely believable further claim to have mastered the low Venetian dialect used in the people's theatre in another week by reading *I promessi Sposi* and Dante (p. 263). Finally there is his admission in the Foreword to Volume II that even his fabulous memory 'was often vitiated by time' and had begun 'to colour incidents dramatically'. 'For example,' he continues, 'I had been told a story by someone, it lay dormant in me for years; suddenly some striking fact called back the tale and I told it as if I had been present' (p. 223). Philippa Pullar, his most recent biographer, interprets this as a reference to his eye-witness account in Volume I of a rail accident in Wales which actually occurred when Harris was in Ireland.[16]

Further improbabilities reveal themselves by the manner and style in which they are narrated. Harris shows a tendency to turn life into literature and to treat literature as life. His account of his life at boarding school is reminiscent both of a *Boy's Own* annual and of *Tom Brown's Schooldays*. A rebel and outcast at school, he leads a fags' revolt which results in his being viciously beaten by Jones, the school bully who is modelled on that archetypal bully, Flashman. Having born the beating 'without a groan', Harris went off to plot revenge. In true boys' adventure story tradition he is taught to box by his elder brother and a policeman friend during the holidays. On his first day back he slaps Jones to provoke a staged fight in which he duly reduces Jones's face to one big bruise and finally knocks him to the ground. When Jones eventually reappears after days spent in a darkened room, Harris magnanimously shakes hands with him just like any good sport in boys' fiction would do. This schoolboy prodigy takes off for New York at 16. On his second day there he joins a bootblack and with his first customer completes two shoes in the time it takes the professional to do one. Next morning he goes to work underwater in a compression chamber helping to construct the foundations of Brooklyn Bridge. Despite working in icy water at tremendous air pressure in a temperature of 180 degrees Fahrenheit (or so he claims), on his shift he and a Swede together 'did more than the four others' (p. 68). Always there is this compulsion to excel at everything, to be the kind of superman found only in myth and fiction.

It is perhaps significant that Harris had already incorporated his month's stint of work on Brooklyn Bridge into a novel, *The Bomb* (1909). It is common for novelists to be unable to disentangle the original models from the fictionalized characters into which they transformed them. In *Christopher and His Kind* Isherwood admits how impossible it is for him forty years later to disentangle Sally Bowles from her real-life original.[17] Not so Harris. He repeats this dubious process when coming to deal with his year's experiences as a cowboy which he had already turned into a full-length novel, *On the Trail*, by 1878, although it was not published until 1930, after the publication of the first four volumes of his autobiography. The section of Volume I covering this period of his life reads exactly like a dime western and was duly incorporated into a movie, *Cowboy*, in 1962. A natural shot, with an instinctual sense of direction, Harris also becomes a superb horseman in no time, winning Blue Devil, the stock temperamental young mare of amazing speed and stamina, by kindness. 'No one she disliked could mount her, for she fought like a man with her forefeet,' Harris yarns, 'but I never had any difficulty with her, and she saved my life more than once' (p. 88). This is a staple ingredient of the western, and other such incidents follow, despite his declared intention of reserving much of this material for the novel. Bitten by a rattlesnake he bites and burns out the sting without a thought. Again and again he appears as the naturally gifted hero, saving part of the herd from the great Chicago fire, becoming a joint partner of the outfit, rustling cattle from Mexico, holding off hordes of Indians, riding through Indian territory for help on Blue Devil, and witnessing the death in a bar-room brawl of a fellow cowboy. Harris clearly has difficulties distinguishing between reality and the fantasy he borrowed from books in which he lived so intensely that they became part of his inner life.

It is not just the situations that are clichéd. The style frequently betrays the extent to which Harris is prepared to replace reality with the language of romance and adventure stories. In his Foreword to Volume III he comes closest to a statement of this intent. 'I have always wanted to build Romance in the Heart of Reality, making the incidents of my life an Earthly Pilgrimage' (p. 524), he writes, echoing in the language he uses the degree of stylization which he is prepared to employ in his treatment of the raw material of his life. His account of the entry of Laura, his first love, into his life draws on the staple devices of the romantic novel: 'She came towards me where I was standing by the window and took my breath,' he reports. She 'swam rather than walked', and had 'a mane of chestnut hair brightened with strands of gold' (p. 310). 'I was in love for the first time,' Harris writes after 342 pages, 'on my knees in love, humble for the first time, and reverent in the adoration of true love.' This reads like the description of a religious experience. At a time of growing religious scepticism that may well be how Harris came to see the missed opportunity of finding love in his sexually promiscuous life. The language he occasionally places in the mouths of his mistresses is even more removed from the real world. One extreme example will make the point. Harris's married mistress, about to be dragged away from Lawrence, Kansas, and Harris by her husband, proposes a solution:

'This life holds nothing worth having but love. Let us make love deathless, you and I, going together to death. What do we lose? Nothing! This world is an empty shell! Come with me, love, and we'll meet death together!'
    'Oh, I want to do such a lot of things first,' I exclaimed. 'Death's empire is eternal, but this brief task of life, the adventure of it, the change of it, the huge possibilities of it beckon me. I can't leave it.' (p. 149)

This amalgam of Shakespeare, Milton, Swinburne and other less exalted writers reads like a second-rate melodrama. It is life twice removed from reality by Harris's blatant literary borrowings. Yet to the extent that he preferred to live out his life as if he were a character in some literary fantasy the highly charged language, however remote from actuality, does reflect an inner state of mind. It is both misleading and revealing.

If the reader feels that I have been unduly sceptical I can only suggest that I abandon the self-imposed restraint of looking for internal evidence of truth or falsehood solely within the autobiography and look at the findings of Harris's recent biographer, Philippa Pullar. She casts doubt over the whole cowboy episode, showing that Harris spent the winter of that year in Denver unloading fuel from freight wagons (an incident he omits). His main source of income in Lawrence came from running a butcher's stall prior to enrolling at the university, again something of which he makes no mention. He returned to Europe via the Atlantic, not westwards as he reports in graphic detail. He was teaching at Brighton when he claims to have been accompanying the Russian general Skobeleff on his campaign against the Turks. Most damaging of all is Harris's omission from the autobiography of his first marriage to Florence Adams which lasted less than a year before she died in 1879. This was during his German sojourn when he was purportedly practising 'the merits of chastity' (p. 229). Philippa Pullar explains his later inclination in the autobiography to overdramatize or sometimes invent his extra-marital liaisons by pointing out that he 'was dictating to (and he hoped titillating) a series of young secretaries, many of whom were in love with him and with all of whom he was having affairs'[18] The facts, then, are even further removed from Harris's version of them in his autobiography than internal evidence has indicated. Yet I hope that I have shown how much one can learn from such internal evidence alone.

Unintentional Revelation: Harris and George Moore's *Hail and Farewell*

In order to demonstrate just how much evidence can be found within the text, evidence which enables the reader to make an independent assessment of the kind and degree of truth to be expected from any particular autobiographer, I have necessarily concentrated primarily on negative aspects of Harris's autobiography. Yet in its way it is a fine contribution to the genre, highly readable, full of incident, ambitious in scope, and by no means as overwritten in general as some of the quotations I have cited might suggest. A more positive approach to internal evidence within this and all autobiographies is to look at the way autobiographers inevitably give more of themselves away to the reader than they intended. As Clive James writes in the Preface to his *Unreliable Memoirs*, 'However adroitly I have calculated my intentional revelations, I can be sure that there are enough unintentional ones to give the reader an accurate impression.'[19] This is one of the unique features of the genre, one which adds to its complexity and fascination. Critics have long been aware of the presence of a less conscious subtext in imaginative writing of all kinds. But this subtext assumes special significance in autobiography because it throws light on author, narrator and protagonist alike. So an author's or narrator's attempt to give an 'objective' picture of the protagonist will unconsciously reveal something more subjective about the way the writer sees himself at the time of writing. The reader of autobiography is consequently drawn into making judgements about the nature and personality of the writer in order to arrive at some estimate of the nature and reliability of the text.

The subtext can be especially revealing about the writer's continuing although unconscious dependence on his relationship to his parents. He may well be blind to the true nature of this relationship. At any rate he would normally attempt to censor the childish component in his adult behaviour both in his life and in his account of it.

Harris has an interesting short subsection at the end of the first chapter of *My Life and Loves* titled 'My Father'. In it he characterizes his father as strict (he once beat Harris for listening to smut), bad-tempered, a first-class sailor, mean (he only got drunk when others paid for it), frightening and overdemanding. He made his son run races up the rigging against experienced sailors and expected him to win. He forced Harris on one occasion to go on swimming round his boat to show him off to a Lord of the Admiralty until the frightened young boy was on the point of collapse. Harris comments: 'The memory of my fear made me see that he was always asking me to do too much, and I hated him . . .' (p. 20). What Harris doesn't seem to realize is that he continued to live up to his father's unfair expectations of him for the rest of his life. Harris is a classic case of a child who has unconsciously introjected his father in the course of consciously separating himself from a parent whom he had come to fear.

What form did this process take? Primarily that of self-vindication. His father evidently expected him to excel. His life is at least partly explainable as the reiterated attempt to win his father's approval (in his own mind) by winning out against the odds. His experiences in the States repeatedly follow the same pattern of the young *arriviste* quickly outshining old hands whether they are bootblacks, hotel managers, cowboys or even professors. Professor Byron Smith seems to have appealed to Harris as the substitute good father. Yet, just as his father drank, so Smith suffers from wet dreams, and Harris reaches an apotheosis of self-vindication when he successfully, if only temporarily, stops their occurrence with the painful application of a whip-cord. Not content with this physiological victory, Harris towards the end of his relationship with Smith also achieves intellectual parity with his one-time tutor and mentor in an argument about individualism and Marxism: 'Soon he began congratulating me on my insight, declaring I had written a new chapter in economics.' Harris comments: 'His conversion made me feel that I was at long last his equal as a thinker' (p. 165). His entire sojourn in Europe as a scholar is at the bidding of Smith and again it is undertaken 'as a sort of consecration' (p. 185), an act of filial piety.

By a remarkable coincidence Harris spent several months before his period as a wandering European scholar at his father's home in Wales. There he claims to have become reconciled with him, especially after being nursed by the old man when the son had drunk poison by mistake. The whole incident is highly suspect, since Harris claims to have drunk sixty grains of belladonna fasting when one grain was considered fatal. Convinced he was dying, Harris has only one regret, a regret that had driven him in the past and that was to drive him on to further feats of heroism in the future: 'I wished that I could have had time to prove myself and show what was in me!' (p. 203). Throughout one critical night his father lay in extreme discomfort by his side holding his hand. 'From that time to the end of his noble and unselfish life some twenty-five years later, I had only praise and admiration for him' (p. 205), Harris writes. The entire incident sounds like a case of wish–fulfilment. By enlisting the help of Smith the bad father has been converted into a benevolent one, first figuratively and then literally. The need to prove himself worthy by excelling has been allayed. Or has it? Harris went on to earn further laurels in German academe and in the worlds of British and American journalism and letters. But he was writing Volume I after he had achieved what successes he had, when he was once more in financial straits and by no means an assured literary success. So the writing of his autobiography became itself an act of self-justification. Through it he could once more quieten that parental voice within him urging him yet again to outdo his contemporaries. Author, narrator and protagonist interact to placate those inescapable paternal expectations in this final *magnum opus*.

If his attitude to the various women in his life is partly conditioned by his need to excel implanted by his father, it may also be partly the result of the loss of his mother when he was not quite

4 years old. Whether or not he actually underwent the traumatic experience of discovering that she was dead on going to kiss her goodnight as he claims, he made of her memory an image of idealized womanhood. During his stay in Ireland during 1881 he visited her grave and reports the judgement of a local minister who had known her that she 'was a saint of the sweetest disposition and very good-looking', an impression which his 'own childish recollection corroborated' (pp. 289–90). Compared to this sanctified image of beauty and goodness, all his mistresses are found wanting. Invariably they are physically flawed in some way. Even in the case of Laura, the great love of his life, he notes that 'the oval of her face was a little round', and 'her fingers were spatulate and ugly'. (p. 342). As was the case in his relationship with his father, Harris seems unaware of how much of his behaviour towards women is the product of this idealization of the mother he was deprived of so early in life. Without intending it, he has provided his readers with evidence by means of which it is possible to achieve an understanding of him that goes well beyond his conscious intentions. One could argue that at an unconscious level he wished to impart this information. The presence of such unintended insights is usually a characteristic of the more personal autobiography on which I am concentrating. The subtext helps the reader to assess for himself the extent to which the autobiographer is in control of his material. It also sets up a rewarding dialectic between intended and unintended levels of truth.

A particularly interesting example of this kind of dialectic occurs in George Moore's autobiographical trilogy, *Hail and Farewell*, consisting of *Ave* (1911), *Salve* (1912) and *Vale* (1914). The three volumes span the years 1899 to 1911. In his late forties Moore became so excited by the first stirrings of the Irish literary renaissance that he returned to his native country to make his home in Dublin for the years 1901 to 1911. During this decade he grew increasingly disillusioned with Ireland and the Irish. Finally he returned to London where he based himself for the rest of his life, timing his move to avoid being in Dublin when the first volume of his autobiography was published late in 1911. During his ten-year sojourn he involved himself with his cousin Edward Martyn and W. B. Yeats in establishing the Irish Literary Theatre, and became an ardent advocate for the Gaelic League's campaign to revive the Irish language. Gradually convinced of the hopelessness of arousing his countrymen to a new awareness so long as they remained cowed by a narrow-minded, philistine clergy, Moore resolved to spend his energy on writing his autobiography, a book intended to save Ireland from its enslavement to Catholicism.

At least those are his declared motives and actions. But there is some confusion between what happened and what he recorded as happening. His growing disillusionment with aspects of Irish life began as early as his first year there with the critics' bigoted reception of the production of *Diarmuid and Grania*, a play he wrote jointly with Yeats. The next year his stories about Irish life, subsequently revised and collected in *The Untilled Field* (1903), already showed a distinct anti-clerical bias. By 1906, only halfway through his Irish decade, Moore was planning to write *Hail and Farewell*. He was at work on it from 1907 to 1913. One of the intriguing questions arising from these dates is at what point Moore ceased to live life for its own sake and began living it for the book's sake. There is so much teleological hindsight in the earlier two volumes that it becomes impossible to disentangle the causes of Moore's disillusionment from the end result—his departure into a second exile from his homeland. By the third volume he is perfectly frank about his decision to leave Ireland 'for the sake of the book'[20] (p. 643). But what is one to make of his conviction, voiced near the beginning of the first volume, that 'an Irishman must fly from Ireland if he would be himself' (p. 56)? Why on his first return to Ireland is he already prepared to find it small as a pig's back, which according to legend it appears to be to its enemies? In the event half the coastline appears like this to Moore on his arrival while the other half seems, as according to legend it does to its friends, like a land of extraordinary enchantment. But the dichotomy set up here is one prepared by its alienated author who already knows that for him Ireland had become no 'bigger than a priest's back' (p. 116). One of the rewards the autobiography offers him is being able to make comparisons between pigs and priests, an analogy which he could not make overtly in the Irish society in which he moved.

Most readers and critics appear to have accepted at face value Moore's claim that *Hail and Farewell* was his 'work of liberation', 'liberation from ritual and priests, a book of precept and example, a turning-point in Ireland's destiny' (p. 643). In an introduction he wrote for a revised edition of the autobiography published in 1925 he claimed that in this book Nature undertook the entire composition: 'Every episode and every character was a gift from Nature, even the subject itself' (p. 51). Far from being its author, he continues, he was nothing more than Nature's secretary. Moore obviously has a heavy personal investment in this unusually impersonal view of his part in his own autobiography. Why is he so committed to this determinist view of what is in fact a carefully constructed account of his Irish period? Why does he insist that this is no ordinary book of reminiscences but a 'sacred book' and that he is no mere autobiographer but Ireland's new prophet destined to free it from the spell which has held it paralysed since the arrival of Christianity?

It is extremely difficult to disentangle the truly personal material from its literary and thematic context. Not only is *Hail and Farewell* extremely carefully composed but it is cast in an elusive mock-heroic mould. Moore's use of a mock-epic tone is partly an ironic comment on the state of Ireland as he found it. It also enables him to view with ironic humour his closest Irish friends – George Russell (Æ), Edward Martyn, W. B. Yeats, Lady Gregory – and, above all, himself. Posing as the mock hero of his would-be epic, Moore is in a particularly well-placed position to conceal his innermost needs and desires during this period of his life. And yet one cannot help detecting behind those lofty sentiments delivered in such a guardedly ironic tone a strong personal investment in interpreting the situation in the way he does. His Irish friends are not just well-known individuals, they are representative figures, 'not personalities', he wrote after the appearance of *Ave*, 'but human types'. 'Edward Martyn, for instance,' he goes on, 'is as typical of Ireland as anything can be; he seems to reflect the Irish landscape, the Catholic landscape.' He feels compelled to expose Martyn, just as he does Yeats ('a type of the literary fop') and Gill ('a fine example of the posthumous intelligence')[21] and, most painfully, his brother Maurice (whose Catholicism represented 'Ireland in essence' (p. 408)) because they represent alternative postures to art and life that he has already rejected.

Moore sees himself too as representative. He stands for the independence of art from restrictions of all kind. He is a late middle-aged Stephen Daedalus flying the same nets of prejudice.

But his reimmersion in Irish life coincides with a more private crisis in his own life, one which he insisted on recounting at the climax of his third volume despite appeals not to. This concerned his growing sexual impotence. Unable to meet the demands of Stella (Clara Christian), the mistress whom he brought with him to Ireland in 1901, he felt forced to give up the relationship in 1904 and stand by helplessly when she married the City Architect of Dublin the following year. She died in childbirth in 1906. The fact that Moore had meantime renewed his acquaintanceship with an earlier admirer, Lady Cunard, had the unexpected effect of reinforcing his conviction that his days of sexual philandering were drawing to a close. But art could still celebrate the past life of the flesh. More than that. It could justify his past life dedicated to the pursuit of women. Beneath his ironic self-criticism lay self-justification. In art lay the same freedom that he had pursued all his life: 'I have come into the most impersonal country in the world to preach persona-lity – personal love and personal religion, personal art, personality for all except God' (p. 609). Behind his mask of the mad prophet or holy fool lies Moore's urgent personal need to give meaning to this crucial new turn in his life.

The writing of *Hail and Farewell* reconciled Moore to the role he was to occupy for the rest of his life, that of the experienced literary master who could draw on an unusually varied past life for the raw material of his later fiction. But simultaneously the autobiographi-cal trilogy enabled him to re-enact and in the process to reconcile himself to the humiliations he suffered during childhood and adolescence. It is no coincidence that the third volume, ostensibly describing Moore in his fifties, devotes the first five chapters to reviewing his earlier years, nor that the volume should end with his return visit to Moore Hall, the ancestral home where his younger brother's family was then living while maintaining the Roman Catholic tradition in which both brothers had been brought up by their parents. The last time he had been there was sixteen years earlier for his mother's funeral in 1895. Moore Hall is made to epitomize Moore's long-held vision of Ireland as a 'ruined country . . . wilderness and weed' (p. 192), 'a god demanding human sacrifices', 'Ireland being a fatal disease' (p. 213). Ireland is 'the terrible Cathleen ni Houlihan', the bitch-mother-goddess who tempts her children into quiescence and forgetfulness. Ireland is constantly alluded to in the autobiography as an all-demanding mother. So is the Roman Catholic Church, Mother Church.

In confronting and finally rejecting these twin mother figures in favour of a free life devoted to art and the love it celebrates Moore is led to re-enact key scenes from his youth. One of the few clear glimpses we are given of his mother in the autobiography shows her laughing derisively at him for taking seriously the parish priest's suggestion that he should give up hunting in favour of an education in the classics. The tragedy of it was, Moore explains, that he 'accepted her casual point of view without consideration, carrying it almost at once into reality, playing truant instead of going to my Latin lesson' (p. 197). A consequence of his accepting his mother's low opinion of his intellectual abilities was that he only narrowly escaped being forced into the army by his father. Saved by his father's death, he left the family home at the age of 21 to educate himself in Paris. Moore's biographer, Joseph Hone, shows that there was no bitter rupture. But he firmly removed himself from all close connections with his family in order to turn himself into an artist.[22] His return to Ireland in late middle life and his almost neurotic avoidance of Moore Hall until he was on the point of a second and permanent parting with his mother country symbolically re-enact and reinforce his earlier break with his widowed mother. Underlying his literary and political alliances and ruptures is a more personal, if less conscious, set of motives, a need to relive his Irish childhood in adult guise.

In a revealing passage near the beginning of Chapter III in *Ave* Moore recollects an occasion in childhood when to break the monotony of infancy he stripped off his clothes and ran naked ahead of his nurse to her embarrassment. He continues: 'Was this visit to Ireland anything more than a desire to break the monotony of my life by stripping myself of my clothes and running ahead a naked Gael screaming Brian Boru?' (p. 111). Here for a moment the subtext becomes overt. The child impelling the adult reveals itself in all its childishness. The parallel between past and present is never far from the surface throughout the autobiography and is overwhelmingly in evidence in the finale where the past possesses him so strongly that it eclipses the present again and again. It is in Moore Hall that he finally realizes that Catholicism and Protestantism are 'states of mind'. 'Every man of worth', he consoles himself, 'chooses a religion for himself', just as he had chosen to commit himself to art between 18 and 21. So that his farewell to Ireland is not just a farewell to his mother country and to his mother church; it is also a re-enactment of his parting from his own mother and all she stood for thirty-four years earlier. The autobiography ends with Moore adapting Catullus' words of mourning when he travelled to cremate his brother's body. In place of 'brother' Moore writes: 'ATQUE IN PERPETUUM, MATER, AVE ATQUE VALE' ('And for ever, Mother, hail and farewell'). He has returned to his bigoted mother country only once more to turn his back on it (as he had done on his mother before), this time conclusively. His self-imposed exile has been vindicated by his dedication to a life of art. His subsequent success as an artist is celebrated and epitomized by this most artful of autobiographies.

Some readers of *Hail and Farewell* might well dispute my particular interpretation of the subtext. One of the features of a subtext is its elusive quality. It requires the reader to tease it out on the basis of fragments of evidence unintentionally incorporated in the text by the author. Nevertheless it is a key element in any autobiography. Autobiographical exegesis depends heavily on it because it offers a second perspective from which a reader can better distinguish the particular prejudices or blind spots in the autobiographer. It is one of the genre's major attractions. Any reasonably undefended autobiographer knows that he is likely to tell his readers more about himself than he intends. Any attentive reader of such an autobiography is constantly on the look-out for clues to this hidden content. It performs a dual function. It reveals a secret area of truth that its author never consciously intended to uncover and it helps to correct the 'official' ego-oriented version of the truth. It is the nearest autobiography comes to expressing the unconscious self. The subtext assumes especial significance in post-Freudian autobiography, because the modern writer is more aware of the unconscious exposure to which the act of writing about himself necessarily commits him. This has tended to produce a high degree of self-reflection and internal tension in modern autobiographies. To this extent they gain over their predecessors by more fully exploiting what has always been a major potential in the genre.

3

Fact and Fiction

I have already offered some examples of how the most factual of autobiographers are forced to fall back on some of the staple techniques of the novelist. Davies drew heavily on the tradition of the adventure story and even Orwell sometimes employed purple prose reminiscent of the *Arabian Nights*. Frank Harris was at times wholly taken over by the various literary genres from which he was borrowing, while George Moore brought to his autobiography all the skills of construction which he had acquired as a novelist. At the same time I have shown that the most novelistic of autobiographers nevertheless provide their readers with internal means of verifying the accuracy of their accounts of themselves. Is this, then, the only distinction between the two genres, fiction and autobiography? Is it even possible to distinguish between autobiographical fiction (such as *Sons and Lovers*) and autobiography, especially when autobiography is confined, as it is in this book, to that branch of it which aspires to the status of a literary form?

It is perhaps worth recalling, in the first place, the way in which fiction modelled itself in its infancy on autobiography. The first critic of the genre, Anna Burr, showed how the entire school of romance fiction dealing with crime and detection originated in a group of memoirs by French *agents de sûreté* of whom the best known was Vidocq.[1] According to Burr, Vidocq's *Mémoires* (1828–9) gave rise to *Les Misérables*, Balzac's novels, Dickens's *Great Expectations*, and Poe's and Conan Doyle's fiction. In England autobiography was firmly established as a mode before Defoe and then Richardson set out to free fiction from its bondage to the artificiality of the romance tradition. In *Autobiography in Seventeenth Century England* Dean Ebner argues that 'seventeenth-century autobiography . . . was the eighteenth-century novel in potential existence'.[2] Defoe pretended to be merely editing the personal memoirs of Moll Flanders and Robinson Crusoe in order to win from his public the credence they were used to giving to autobiographies. Even that most sophisticated of eighteenth-century novelists, Laurence Sterne, modelled *Tristram Shandy*, yet another sham autobiography, on seventeenth-century autobiographies such as Lord Herbert of Cherbury's.

It is such a commonplace to believe that subjective autobiography acquired most of its literary characteristics by appropriations from the novel that it is salutary to be reminded that both genres began being widely practised in England in the last quarter of the sixteenth century.[3] The novel became the more popular of the two only once it had absorbed many of the features of autobiography – its specificity, its 'realism', its looser structure and more colloquial prose style. By the nineteenth century, as Phyllis Grosskurth has argued, it seems as if only in the novel could the Victorians write an autobiography which was also a true confessional.[4] This explains why to modern eyes autobiography appeared to come into its own in the twentieth century only by learning from the nineteenth-century novel many of its narrative procedures. Seen in a wider historical perspective it had to acquire some of the features, like psychological particularity, which the novel had originally learnt from earlier autobiography. During the course of the twentieth century autobiography has ceased to be seen as a branch of historiography, which is how it was treated in the nineteenth century. Today, under the influence of Structuralism and its aftermath, it is widely viewed as all but indistinguishable from fiction.[5] In either case the text becomes autonomous.

The confusion between the two genres has been exploited by modern novelists for their own autobiographical purposes. There is the much-repeated argument, for instance, voiced by Mauriac, that 'every great work of fiction is simply an interior life in novel form'.[6] Literary autobiographers repeatedly refer the reader to books or stories in which they have described true incidents in their lives under the guise of fiction. Normally this approach is found among autobiographers more concerned to correct the record than to produce a literary work of art. Thus A. E. Coppard in the opening page of his autobiography wittily demonstrates the difficulties of writing a book about a life he has already described indirectly in his tales. The autobiography, he explains, 'may – but I know it will! – turn out to be just another work of fiction under the guise of autobiography instead of the deeper autobiography which is after all the principal source of fiction'.[7] The result, compared to his fiction, is a disappointment, as his definition leads one to expect. Any autobiography written to this prescription is likely to prove inferior to the imaginative writing into which an author has directed his full creative talent. It is interesting that at the end of his autobiography Coppard has to admit that all his attempts to confuse the genres have been a protective delusion: 'My autobiography has no more pure fiction in it than my fiction has pure autobiography.'[8] He fallaciously equates the genre's demand for truth with artistic inhibition and uses this as an excuse for not matching the best of his fiction. This is similar to the argument Virginia Woolf offered in her essay on biography to show that biography is intrinsically less creative than fiction.[9] Coppard (and others like him) has confused autobiography with biography which is undeniably limited by its necessity to interpret a range of given facts. The autobiographer, however, is free to omit whatever facts he chooses to, as Jung, for instance, does with abandon in *Memories, Dreams, Reflections.*

One highly original response to these dilemmas facing the novelist attempting to write his autobiography is that of J. D. Beresford. Like many of his fellow novelists, he is convinced that all his fiction is semi-autobiographical. 'Even in my worst books . . . there has always been some idea that represents my personal struggle with the things that seem to be.'[10] So in *Writing Aloud* he describes how a novel (which never got written) came to be planned over a period of four or five years. In doing so he contrives to describe a good deal of his own life and beliefs. The book is a portrayal of 'me and my method', and as such is 'a reasonably honest piece of autobiography'.[11] As the outline of the proposed novel hardens, so Beresford finds it become increasingly a reflection of his own experiences and ideas. Finally he realizes that his hero's life has come to reflect his perception of the recent history of humanity. 'The truth is', he concludes, 'that my single pleasure is in the continual retelling of the story of my own intellectual and spiritual life.'[12] The book offers a fascinating insight into the workings of a novelist's imagination and the manner in which his own life enters that of his fiction. But the embryonic novel he is planning throughout the book lacks vitality, which is why, one suspects, he sacrificed it to this pseudo-autobiography. In this case it is the fictional element which is inferior while Beresford's creative energy has gone into the autobiographical interpretations of what lies behind the fiction in embryo.

Is it even possible to combine, let alone merge the two genres of fiction and autobiography? In his autobiography George Moore was careful not to invent (at least intentionally) incidents, although he felt free to rearrange them where the theme demanded it. As he saw it, if he 'were to introduce a thread of fiction into this narrative, the weft would be torn asunder' (p. 288). Or in other words, the autobiographer's pact with the reader would be broken. In *L'âge d'homme* (*Manhood*), the first of his many books of autobiography written in 1939, Michel Leiris offered a detailed defence of this purist approach to writing autobiography in an essay, 'The Autobiographer as Torero', first included in the 1946 edition, which is attached to the 1968 edition of *Manhood* as a foreword. In his autobiography, he writes, he aspires to the graceful courage of the matador who turns personal danger into an art. Setting out to endanger himself by exposing his most humiliating deficiencies in his autobiography to the horns of the reader he comes to realize that nevertheless 'every confession contains a desire to be absolved'. To use the novelist's art is to seduce the reader into absolving the writer, 'to limit – in any case – the scandal by giving it an aesthetic form'. Leiris therefore imposes rigid anti-fictional rules on his own autobiographical writing:

From the strictly aesthetic point of view, it was a question of condensing, in the almost raw state, a group of facts and images which I refused to exploit by letting my imagination work upon them; in other words: the negation of a novel. To reject all fable, to admit as materials only actual facts (and not only probable facts, as in the classical novel), nothing but these facts and all these facts, was the rule I had imposed upon myself.[13]

But he fails to exclude his imaginative and interpretative powers from the autobiographical act. Just the space given to and juxtaposition of the facts and images he selects create a pattern and a meaning of their own. *Manhood* is largely structured on two dominant images of women that have governed his life, those of the sacrificial Lucrece and Cranach's vengeful Judith. As if in recognition of this contradiction, Leiris opens his next book of autobiography, *Biffures* (*Erasures* or *Bifurcations*, 1948) with: 'Il était une fois . . .' ('Once upon a time . . .'). From the start he is warning the reader of what he has come to see as the inescapable fictionality of autobiographical truth. *Biffures*, among other things, is an exposé of the fictional component of fantasy present in his verbal constructions. Yet he never fully gave up the search for rules by which to play the game of autobiography. In *Fibrilles* (1966), the third of his four volumes of autobiography to which he gives the collective name of *The Rule of the Game*, he once more tries to spell out the rules necessary for autobiographical authenticity. Finally in *Frêle Bruit* (1976), his fourth volume, he implicitly hands the fragmentary materials of his life over to the reader to make what sense or unity of it he can. Let the reader, he implies, be responsible for the inescapable element of fictionalization.

One English novelist has responded to this classic autobiographical dilemma by abandoning any attempt to disentangle pure autobiography from pure fiction. In *Raw Material* (1972, 1974, 1978) Alan Sillitoe interweaves sections exploring his past with a provocative series of meditations on the nature of truth in life and in books. He refuses to call it an autobiography 'because everything written is fiction, even non-fiction' (p. 9).[14] He leaves out most of the life history one expects in a conventional autobiography on the Shavian grounds that much of this material has already been used in his fiction to date. Instead he concentrates on describing his grandparents, because, he writes, 'Those whom I knew so well are part of my corporate identity' (p. 67). But the truth is by nature elusive. The artist's desire to convince other people of the truth, he asserts like Leiris, involves him in lying. As A. E. Coppard wrote in his autobiography, 'What is important in the art of fiction is not the truth itself, but the sensation of truth.'[15] So, Sillitoe insists, to 'refuse the responsibility of a lie means pushing art out of the way, for it is only possible to create art when seeking to make raw truth believable' (p. 63). Sillitoe here subscribes to the often repeated myth that truth can be so much stranger than fiction that if it were presented in its raw state it would not be believed.

The autobiographer's dilemma, then, according to Sillitoe, is the need to choose between telling the truth and convincing the reader that he is telling the truth. Like Leiris, too, Sillitoe foregrounds this dilemma, in his case by interlarding his historical family narrative with a long series of meditations on the nature of truth. Again and again he seems able to express himself only in paradox, offering the reader a multiplicity of carefully phrased, mutually contradictory aphorisms. The past is fiction, he asserts. The present is illusion. 'Yet out of this past which has become fiction I fish for the truth' (pp. 169– 70). Afraid that if he were to reach even some of the truth he would be reduced to silence, he consoles himself by reflecting (much like Chesterton) that, 'After a while even the most stunning truth no longer shines or intimidates because of the familiarity it has meanwhile gained. One then denies it, and looks for it once more' (p. 20). This suggests that like Leiris one could spend a lifetime writing different versions of one's life story, because of the relative nature of autobiographical truth. The most interesting point that emerges from Sillitoe's attempt to obfuscate the distinction between fact and fiction is his identification of truth with the shock of the new. One way of encountering truth anew is by the exploratory process of writing about one's past before familiarization sets in. 'All truth is fiction' (p. 189), Sillitoe perversely decides, and ends his autobiographical novel or fictional autobiography with a question, since only the open-ended nature of questions partake of the divinity of truthfulness.

While it is perfectly legitimate to undermine some of the common assumptions about the distinctions between fiction and autobio- graphy, I am not convinced that the autobiographer's pact with the reader can be reasoned away. The intention to tell the truth about oneself as best one can differentiates the expectations of the reader of autobiography from the reader of fiction. It also affects the writer of an autobiography. As Shumaker observes, 'Whereas the novelist, within limits, may "create" persons and situations, the autobiographer can only re-create.'[16] The autobiographer can shape , dramatize or stylize his material, but he cannot knowingly invent it. Not that his material has to be confined to facts, as we have already seen. It can include memories, however inaccurate, reported fact, true or false, and fantasies, dreams and myths all of which avoid the true/false dichotomy. Autobiographies are further distinguished by the fact that the protagonist, narrator and author share the same name. This means that all the staple ingredients of narration – theme, form, characterization, style, imagery – cast light

on the subject of the book and are integral to the autobiographical
process of self-portrayal. As Motlu Blasing puts it, 'The "I" in its
self-consciousness constitutes at the same time the historical
subject, the shaping form, and the personalized style of autobiography.'[17] In other words the autobiographer's use of literary
techniques normally (if incorrectly) associated with the novel does
not necessarily undermine the truthfulness of the writing, as Sillitoe
assumes it must. The attempt to win the reader's interest by use of
an ingenious structural design or by resort to caricature simultaneously informs the reader about the nature of the author who
shares an identity with his narrator and his protagonist, so that its
literary features become part of the overall autobiographical
strategy.

Form: Conrad's *A Personal Record*

Form has always been considered a distinguishing feature of fiction.
Successful autobiographers are equally reliant on form which helps
to differentiate them from such associated genres as diaries, letters,
res gestae and reminiscences, all of which are normally too
controlled by external events to be shaped into a subjectively
meaningful design. Most autobiographers are convinced that life is
largely shapeless, that, as H. G. Wells complains, their life's 'story
has no plot'.[18] Unlike Wells some plead this as their excuse for the
formlessness of their autobiographies. Richard Aldington, A. E.
Coppard, William Plomer and Goronwy Rees all argue in their
different ways that the imposition of form on the material of their
lives would falsify the episodic nature of them. Goronwy Rees
consciously discards chronological continuity in his autobiography,
A Bundle of Sensations (1961), because he feels that there is no
continuity of personality. Then all autobiographers are faced with
the fact that they can have no conscious knowledge of their birth or
their death. Geoffrey Grigson attacks the way most autobiographers begin with 'I was born . . .' when actually 'I' was not. He
asks, 'Why begin with an event outside consciousness which one
only accepts on hearsay?' But he fails to make use of this insight.
Instead he succumbs to convention by continuing: 'I was born all
the same in 1905 . . .'[19] Richard Aldington, Elizabeth Bowen and
Graham Greene all draw attention to the arbitrary nature of an

autobiography's ending. Unable to end with his death, Graham Greene chooses to bring his autobiography (appropriately called *A Sort of Life*) to a sort of death with the years of failure that succeeded the publication of his first well-received novel. In *Life for Life's Sake* (1941) Aldington sees that a sense of form would require the book to end with his departure for the United States in 1935. Nevertheless he feels obliged to 'round the story out a little',[20] supposedly for the reader's sake, by continuing his life history up to 1939–40. Elizabeth Bowen confronts the reader directly with the enigma of autobiographical form, which is particularly acute in her case since she has chosen to confine herself to the first seven years of her life. 'Shall I write "The End" ', she finishes by asking 'to a book which is about the essence of a beginning?'[21] All autobiographical conclusions have to be created by writers out of the continuity of their lives. What they create is the artistic illusion of conclusiveness.

Even Leiris, who is so suspicious of the well-constructed autobiography, concludes that it is permissible to seek for a form that contributes to the autobiographer's overriding aim of self-exposure, just as the *corrida* 'is a technique of combat and at the same time a ritual',[22] one which enables the public to participate in the matador's brush with death. Leiris is here distinguishing between the external imposition of literary form and what Georg Misch has called 'inner form' which 'gives the recollected material form from within'.[23] George Moore's *Hail and Farewell* is an excellent example of the use of inner form where even the titles of his three volumes reflect accurately his changing response to his ten-year return to Ireland. *Ave* enthusiastically hails the Irish literary renaissance to which he contributes. *Salve* shows his growing concern for the health of his native country. *Vale* painfully bids farewell to it. Inner form structurally reflects the theme of the book, the theme being concerned with the meaning which the reflecting consciousness of the autobiographer makes of his life in the process of, not creation, but re-creation. One might say that inner form evolves out of the autobiographer's urge to make sense of his life. Form in this sense represents the degree to which an autobiographer has successfully discovered a shape or meaning to the life about which he is writing.

An extreme instance of the use of inner form in autobiography occurs in Joseph Conrad's autobiography, *A Personal Record: Some Reminiscences*. For some reason this short book has rarely received the attention due to one so skilfully structured. This may be due partly to the circumstances of its composition. It was originally undertaken at Ford Madox Hueffer's prompting to appear in seven consecutive issues of Hueffer's influential periodical, *The English Review*, between December 1908 and June 1909. It was dictated to a secretary, transcribed, then revised by Conrad. After seven issues Conrad abruptly curtailed his reminiscences, first claiming that gout had prevented him from completing the eighth instalment, then breaking with Hueffer whose financial mismanagement of *The English Review* had caused him to lose editorial control of it.[24]

Hueffer apparently wrote to Conrad protesting about his sudden discontinuaton of the reminiscences. Conrad's reply reveals the fact that by this stage he had discovered in the seven instalments an inner form which he was determined not to lose. Protesting at Hueffer's reproach that he had left the reminiscences in a 'ragged condition', Conrad wrote:

It is another instalment which would make the thing *ragged*! It would have to begin another period and another phase. On a dispassionate view I see it so clearly that nothing on earth would induce me to spoil the thing as it now stands by an irrelevant single instalment.[25]

The same letter provides convincing evidence that Conrad's deprecating remarks about the reminiscences to H. G. Wells (to whom he wrote of them as 'megalomaniac's stuff' and a 'silly enterprise')[26] represent his characteristic nervousness at exposing his personal life and not a considered opinion of their literary merit. The latter comes in his letter to Hueffer:

It expresses perfectly my purpose of treating the literary life and the sea life on parallel lines, with a running reference to my early life. It treats of the inception of my first book and of my first contact (psychologically and *de facto*) with the sea. It begins practically with the first words of appreciation of my writing I ever heard and ends with the first words ever addressed to me personally in the English tongue. And actually the very phrase ending the seventh instalment is to my mind an excellent terminal, a perfect pause carrying out the spirit of the work . . . .[27]

As for Conrad's use of dictation, if anything it enabled him to stay closer to the spirit of autobiography than his natural reticence would have allowed him to do in the more deliberative act of writing. In 'A Familiar Preface' (written for the first publication of the reminiscences in book form in 1912) he admits to having 'a positive horror of losing even for a moment . . . full possession of myself'.[28] The act of dictation allowed the associative functions of Conrad's mind fuller play and so helped indirectly overcome this inhibition. His sure sense of form led him to override conventional chronological sequence, opening the book at a point in time later than that on which he closes it. In fact *A Personal Record* is an excellent example of one unusual application of the theory of fictional impressionism that Hueffer and Conrad developed between them. Named after the French Impressionist painters, it replaced chronological with psychological realism in the novel. Life, they argued, is experienced as a series of impressions, not a chronological sequence of events. Reality has no independent existence outside the mind of the perceiver. Hence their use of a narrator such as Conrad's Marlow or Hueffer's Dowell, through whose impressionistic consciousnesses all the events that occur are filtered. In the case of his autobiography, Conrad splits himself between the present-day narrator behind whom the author takes cover and his various past selves, the protagonists of his narrative.

Central to the conception of impressionism is the technique of fragmented time. As Ford Madox Ford (by then) explained:

We wanted the Reader . . . to be hypnotized into thinking . . . that he was listening to a simple and in no way brilliant narrator who was telling – not writing – a true story . . . And it is in that way that life really presents itself to us . . . dallying backwards and forwards, now in 1890, now in 1869, in 1902 – and then again in 1869 – as forgotten episodes came up in the minds of simple narrators.[29]

Because of the nature of the genre Conrad could not adopt the pose of too simple a narrator, although he does conceal a good deal of his artistry behind a façade of casual reminiscence. In place of chronology Conrad substituted a sequence which reflected the associative way in which his mind worked as his narrator thought out loud about his past life. *A Personal Record* is a prolonged meditation on the past which is given shape by the present-day narrator with his need to discern patterns of continuity with his past. These patterns necessarily illuminate the psychology of the pattern maker, and from them, as Conrad hoped, emerges 'the vision of a personality' (p. xxiii). As he realized, his personality was unusually disparate, combining a land-locked Polish childhood with an early adult life spent at sea and a late career as an English novelist – what his biographer, Frederick Karl, has called 'The Three Lives'.

The structure of the first section alone serves to illustrate how throughout the book Conrad's impressionist technique throws emphasis on the psychological unity of the protagonist's personality at the expense of chronology. The autobiography as a whole contrasts the period 1889–94 during which Conrad wrote his first novel, *Almayer's Folly*, with the years 1873–4 when he fought his ultimately successful battle with his uncle to be allowed to go to sea. Section 1 opens in 1893 with Conrad beginning the tenth chapter of his novel on board his last steamship. In one sense the opening links the twin themes of the book, writing and seafaring, or art and life, at the very moment when Conrad was on the point of abandoning the sea for the literary life. The figurative way in which he expresses this simultaneously reveals the fact that one of the structuring devices in the book is the recurrent motif of a little death followed by a form of rebirth: 'The sun of my sea-going was setting too, even as I wrote the words expressing the impatience of passionate youth bent on its desire' (p. 4). As a passionate youth Conrad himself had to die out of his Polish nationalist upbringing to be reborn as a professional seaman, just as he had to kill off his restless life as a sailor to begin writing about the passionate youth of his first heroine. Earlier still his childhood was punctuated by the deaths of his mother when Conrad was 7 and of his patriot father when Conrad was 11, leading to his acquisition of a new father figure and mentor, his Uncle Tadeusz Bobrowski – yet another form of rebirth into a different life to that he had spent with his father in exile.

Within this wider context the first section moves effortlessly back in time, first by a month to November 1893 when he accepted a position on the ship, then to four years before when he began the novel in a London lodging. After a brief return to 1893, he explains how the novel was repeatedly interrupted by long sea journeys, one of which (to the Congo in 1890) takes him back to 1868 when as a 9-year-old boy he put his finger on the blank space in a map of Africa and announced that he would go there when he grew up. Next comes an account of where he was during the writing of Chapters 7 to 9 of the novel, 1890–1. Stuck in the ninth chapter

Conrad received encouragement from an educated passenger on one of his ships in 1892. Still unable to restart the novel, Conrad was summoned back to Poland by his sick uncle in the summer of 1893. This visit in turn makes him recall an earlier visit he had paid his uncle with his mother in 1864 (actually 1863). The section ends with his demonstrating how this evocation of remote childhood memories is not irrelevant to the story of the evolution of his first novel, because it represents 'the still voice of that inexorable past from which his [Conrad's] work of fiction . . . remotely derived' (p. 25). Unity of personality is his theme and this theme generates seemingly effortlessly the inner form of his reminiscences.

In subsequent chapters Conrad allows himself to wander even further back in time to his great-uncle's participation in the Napoleonic retreat from Moscow in 1812, as well as forward in time to the composition of *Nostromo* in 1904 and to the time of writing (1909) at the end of Section 5. He even looks forward to the future when he indulges in an imagined conversation after his death with the real-life original of Almayer. But my synopsis of the structural skeleton of Section 1 should be sufficient to show how chronology is continually disrupted to subserve the psychological requirements of self-portrayal. He can convince the reader of the unity of his disparate lives only by constantly juxtaposing similar facets of his personality dating from different moments in his life. It is for the reader to attend to the inner form, to make the connections necessary to see the subject whole. Conrad calls the book 'a bit of a psychological document' (p. xxii), because in it he presents the evidence in an order which reflects the associative way he remembers life and which consequently gives the reader indirect insight into the author's later personality.

Conrad's impressionistic technique appears so natural partly because it reflects the unforced convictions of a lifetime. He cannot help making the connections on which the inner form of his book depends, since these connections were all made first by the personality which is the subject of his reminiscences, a personality also belonging to the writer of them. Seen this way, his autobiography implicitly endorses the concept of continuing identity between past and present selves, past protagonists and present narrator-cum-author. Again and again Conrad ingeniously demonstrates how the later artist makes use of the same skills acquired by the earlier seaman. For both personae the Board of Trade's demand for sobriety is a necessary qualification:

I will be bold to say that neither at sea nor ashore have I ever lost the sense of responsibility. There is more than one sort of intoxication. Even before the most seductive reveries I have remained mindful of that sobriety of interior life, that asceticism of sentiment, in which alone the naked form of truth, such as one conceives it, such as one feels it, can be rendered without shame (pp. 111–12).

Here is disclosed the autobiographical origin of his literary theory of impressionist indirection. In fact his fear of intoxication and excess is traced much further back, to the exile into which his father's excess sent him, and even to the earlier excesses of his great-uncle, Nicholas B(obrowski) to whom he devotes most of Section 3. Yet at the same time Conrad acknowledges that both the major decisions of his life, to go to sea and to become a novelist, were acts of excess performed in a state of intoxication.

I dare say I am compelled, unconsciously compelled, now to write volume after volume, as in past years I was compelled to go to sea, voyage after voyage. Leaves must follow upon each other as leagues used to follow in the days gone by, on and on to the appointed end, which, being Truth itself, is One – one for all men and for all occupations (p. 18).

Art and life are inescapable in his vision and he proves the point by continuous linguistic confusion between writing and navigating the oceans. Talking of the twenty months during which he wrote *Nostromo*, he describes how he had ' "wrestled with the Lord" for my creation, for the headlands of the coast, for the darkness of the Placid Gulf . . .' (p. 98). Before one realizes one is with Conrad at the helm of his literary craft drawing on his seaman's skills to effect a voyage on paper. He also makes much of the fact that he finally qualified as an *English* ship's captain, that his introduction to the English language came largely from his fellow 'English and Scots seamen . . . who had the last say in the formation of my character' (p. 100). Yet he undercuts the idea that his novel writing in English is primarily the outcome of qualities he acquired at sea by recounting an episode from his boyhood when at the age of eight he read to his father the proofs of his father's translation of Victor Hugo's *Toilers of the Sea*. 'Such', he comments, 'was . . . my first introduction to the sea in literature' (p. 72). Frederick Karl confirms that it was the sea adventures of Hugo, Marryat and James Fenimore Cooper that first aroused Conrad's desire to travel. This circularity of experience, from literature of the sea to life at sea and from life at sea back to his own literature of the sea, gave Conrad the confidence to bend the chronological straight line of his life into circles of timeless personal truth which the inner form of his reminiscences reveals.

To help structure his autobiography Conrad borrowed a further fictional technique that he had jointly developed with Hueffer. This they called *progression d'effet*. What they were seeking was a form of psychological progression in their novels, an accumulation of shocks to the understanding until the total underlying meaning of the work had been laid bare by the end. As Ford Madox Ford explained it, the 'whole novel was to be an exhaustion of aspects, was to proceed to one culmination, to reveal once and for all, in the last sentence, or the penultimate, in the last phrase, or the one before it – the psychological significance of the whole'.[30] Conrad is equally concerned in *A Personal Record* gradually to uncover by a series of revelations (in the form of juxtapositions of chronologically disparate incidents) a unified concept of personality that had been there from the start. His romanticized longing for England and the English is one instance of this. Section 2 introduces this theme by carrying him back to 1873 when at the age of fifteen he went on a walking holiday with a tutor who had been briefed to dissuade him from his intention of going to sea. While Conrad is arguing his case an Englishman passes them by – his 'first contact with British mankind', 'an ardent and fearless traveller', 'the ambassador of my future' (pp. 39–41). Conrad's mind is made up and he successfully withstands his tutor's arguments. Section 4 contains a history of his gradual acquaintanceship with English literature. Section 5 glances forward to his writing of *Nostromo* at the height of his powers as an English novelist. Section 6 reveals how at the age of fifteen to sixteen he had already made up his mind, 'if a seaman, then an English seaman' (p. 122). All this leads to the finale in Section 7, when at the very start of his career at sea he rows out with some French pilots in Marseilles and his 'hand touched, for the first time, the side of an English ship' (p. 134). He is hailed by a deckhand in English – 'the speech of my secret choice, of my future . . . of my very dreams' (p. 136). As he pushes his dinghy off from the English ship, he writes, 'I felt it already throbbing under my open palm' (p. 137). As the ship disappears towards the harbour it raises the Red Ensign, 'the symbolic, protective bit of warm bunting flung wide upon the seas, and destined for so many years to be the only roof over my head' (p. 138).

Seeing that Conrad began the autobiography portraying himself as an experienced ship's officer already writing his first novel in English, this ending hardly comes as a surprise in terms of plot. But it does have a psychological shock value, ending years before the opening pages with an affirmation of his romantic attachment to the English language and English civilization that jolts the reader into a final realization of how unified Conrad's disparate existences have been. This clinching insight into what Conrad calls his 'coherent, justifiable personality both in its origin and in its action' (p. xxiii) successfully vindicates his claim in 'A Familiar Preface' that his 'memoirs put down without any regard for established conventions have not been thrown off without system or purpose' (p. xxiii). The system has been borrowed from his experience as a novelist, but the purpose (the emergence of 'the vision of a personality') has transformed the system to an instrument perfectly adapted to his autobiographical needs. Conrad was well aware of the danger, since spelt out by Leiris and Sillitoe, involved in applying fictional techniques to autobiography. In 'A Familiar Preface' he writes that 'the danger lies in the writer becoming the victim of his own exaggeration, losing the exact notion of sincerity, and in the end coming to despise truth itself as something too cold, too blunt for his purpose' (p. xx). Conrad's application of his fictional theories to his reminiscences contrives to avoid this particular kind of falsehood by rearranging the historical landmarks in his life in an order that reflects the unified personality of the subject through the vision of the writer whose theories are a product of that personality. What could be more natural than that a writer of fiction should give fictional form to the truth about himself? In the case of literary autobiography inner form can be self-validating.

Characterization: Isherwood's *Lions and Shadows*

Roy Pascal has argued that the novel has certain technical advantages over autobiography, one of which is the ability of the novel to change its point of view. A novelist can enter into the minds and thoughts of characters other than those of the protagonist. 'The autobiographer', on the other hand, he asserts, 'can neither get inside other people nor outside himself.'[31] In a strict sense one might equally claim this to be the case of the novelist, who merely pretends to be able to do otherwise. So many novelists have asserted, like J. D. Beresford, that they were 'not one, nor two, but fifty people',[32] most of whom they embody in endless variations in their fictive characters. E. M. Forster defined characters in fiction as pure constructs of the novelist's make-up. 'The novelist', he wrote, 'makes up a number of word-masses roughly describing himself . . . gives them names and sex, assigns them plausible gestures, and causes them to speak by the use of inverted commas . . . These word-masses are his characters.'[33] Otherwise, like the autobiographer, the novelist is compelled to describe characters, frequently modelled on people he knows, from the outside. The temptation of the novelist is to pretend to have inside knowledge where none of us have this in real life. More realistically the genre forces the autobiographer to confine himself to the limits of his own perception. The portraits of others that one finds in autobiographies can be just as successful as any to be found in the same author's works of fiction. One critic, David Thornburn, for instance, has argued that Conrad's portrait of the original model for Almayer in *A Personal Record* 'surpasses anything in *Almayer's Folly* itself'.[34]

What is unique to the genre is that all the portraits of other people in autobiographies contribute to the central self-portrait. Thus the real-life Almayer in Conrad's autobiography is one extended instance of Conrad's own need to believe in the convictions of the inner self, no matter what the outside world thinks. *A Personal Record* is very much concerned to justify Conrad's departure from his country in the face of charges of desertion from his compatriots, just as it is an attempt to prove the inner integrity of his writing in the face of adverse criticism. Almayer's belief in himself, surrounded as he is by a hostile jeering society, offers a pattern of behaviour to which Conrad aspired himself.

Once it is recognized that characters in an autobiography inevitably contribute to a writer's self-portrait, one can understand why so many autobiographers feel free to paint the portraits of those closest to them with all the skills they have acquired in writing fiction. So long as they don't knowingly lie they assume the same freedom to elaborate on, interpret and categorize their friends and relations as they do in their fiction. What is more, if Forster is right that the difference between *Homo sapiens* and *Homo fictus* lies in the fact that *Homo fictus* consists of 'people whose secret lives are visible',[35] then the autobiographer's central self-portrait also paradoxically qualifies as fiction. It is self-evident that all literary portraits are artefacts. Fictional characterization derives from that earlier obsession with typology in human personality first popularized by Theophrastus, each of whose 'Characters' consisted of a human failing followed by an example of it. This original concern with the representative nature of a verbal portrait has continued to haunt the delineation of character in fiction and autobiography to the present day. The impact of the *nouveau roman* with its faceless anonymous heroes has been to re-emphasize the typology underlying all fictional characterization. Whether one accepts Northrop Frye's stock types or Propp's roles from the folk tale, the trend of modern criticism has been towards showing how the reader's response to characters is primarily conditioned by a set of prototypes pre-existing in everyone's mind. The reader searches for signs in the text which will enable him to compare particular characters with their prototypes.

In an autobiography the reader looks at the major characters surrounding the protagonist as influences on, potential expressions of, facets of the protagonist's personality. The skilled autobiographer, conscious of the fact that any major character who is introduced to the book must be shown to contribute to the central portrait, will use the novelist's art to show the character in the role which he or she plays in the protagonist's internal psychodrama. This is the way Frank Harris uses Byron Smith—as the inspiring, intellectual, loving father-figure which Harris needed in order to turn himself into the literary phenomenon that he believed he had become. Similarly George Moore pictures Edward Martyn and his own brother as embodiments of potentialities within himself to sacrifice individual freedom for the safety of social homogeneity. By covertly adding to their self-portraits in this way, autobiographers are referring the art of fictional characterization back to its realistic basis in the writer's self outside which he can never get.

'Character' means 'distinctive mark'. It is surprising how often autobiographers resort to caricature. 'Caricature' derives from *caricare*, 'to load'. By loading or exaggerating the distinctive marks of the major participants in his life's drama, an autobiographer is better able to dramatize the role they have played in his own interior life. Stephen Spender admits to portraying 'types' on occasions on the grounds that 'people tend to become types within certain situations'.[36] John Cowper Powys goes further by caricaturing himself, as if he were one of his own fictional characters: 'A touch of caricature is what we *must* have, if we are to compete, even in this analytical job, with the beautiful madness of Nature.'[37] Of course Nature is mad in his eyes and not necessarily in ours. But what matters is that the portrait should be handled in the artistic manner most appropriate to the portraitist. The use of caricature is common to most comic autobiographies which I will be looking at in the next chapter. But caricature is by no means confined to the comic end of the autobiographical spectrum. It is employed (often humorously) for primarily serious purposes in both of Isherwood's autobiographies, *Lions and Shadows* (1938) and *Christopher and His Kind* (1977). I will confine myself to the former for purposes of illustration.

Isherwood subtitles *Lions and Shadows* 'An Education in the Twenties'. By that, he explains in a preliminary note, 'To the Reader', he means 'to describe the first stages in a lifelong education – the education of a novelist'.[38] That he, Isherwood, happens to be the subject of this sociological analysis is of secondary importance, he goes on, since 'everyone must be his own guinea pig'. Taken at face value this note would suggest that the historical theme of the book is likely to override the central task of self-portrayal. In fact the note is a smoke screen thrown up by the 33-year-old Isherwood to cover his traces. He later confessed that the book is 'factually . . . quite accurate', apart from occasionally 'telescoping similar incidents and situations into each other so as not to weary the reader'.[39] The note is just as evasive in its reference to the characters of his friends:

I have used a novelist's licence in describing my incidents and drawing my characters: 'Chalmers', 'Linsley', 'Cheuret' and 'Weston' are all caricatures: that is why – quite apart from the fear of hurt feelings – I have given them, and nearly everybody else, fictitious names.

Isherwood's attempt to confuse the genres of fiction and autobiography was one way of overcoming the obvious difficulties anyone has in writing an autobiography so early in his life. As a cover-up it was not very successful. More than one reviewer at the time dismissed the preliminary note and discussed the book as pure autobiography. This is what it is if one accepts Isherwood's own definition that true autobiography 'presents a central character to whom all other characters and all events are directly related, and by whose mind all experiences are subjectively judged'.[40] But the need to protect his friends did encourage Isherwood to use the techniques of the novelist with greater freedom than is the case in his much later autobiography, *Christopher and His Kind*, especially in the case of his portrayal of character.

Lions and Shadows covers the years 1921 to 1929, from his entry at the age of 17 to the history sixth form at Repton public school to his departure for Berlin in his twenty-fifth year. His last years at Repton and his life at Cambridge are dominated in the autobiography by the magnetic personality of his friend Chalmers (Edward Upward in real life). Encouraged by their history master to cultivate eccentricity, Chalmers and Isherwood soon adopt the pose of literary conspirators in the midst of an unenlightened community. Gradually they invent a fictive world of their own that they call 'Mortmere', inhabited by a cast of grotesques to which they add by writing half-finished stories about it for one another. These invented characters come to assume almost as much life in the autobiography as those, like Chalmers, which are closely modelled on real people. This is an intentional device on Isherwood's part to illustrate the extent to which Chalmers and, through Chalmers, he had modelled their behaviour on their own fictional fantasies, themselves the product of a common neurosis.

If one selects just one character from Mortmere, Gunball, it is easy to see how he reflects his two authors' predilection in their own lives for surreal fantasy as opposed to the boring reality of lectures and examinations:

Gunball's world was the world of delirium tremens: he saw wonders and horrors all about him, his everyday life was lived amidst two-headed monsters, ghouls, downpours of human blood and eclipses of the sun – and everything he saw he accepted with the most absolute and placid calm (p. 63).

Gunball reminds one of Chalmers's own lurid response to his immersion in Cambridge, 'the country of the dead' (p. 15). His letters to Isherwood from Cambridge are virtually indistinguishable from his Mortmere stories: 'Cambridge is a monster, a blood-supping blasé monster,' he writes to Isherwood. 'It attacks you when you are off your guard, and before you know where you are all poetry and individuality have been drained out of you . . .' (p. 28). Both the fictional Gunball and the real-life Chalmers are rendered as caricatures by Isherwood whose earlier self was becoming unable to distinguish between his inventions and reality.

On their first visit to Cambridge Chalmers warns Isherwood to be on guard against their 'adversaries', the dons, who would be sly, polite, reassuring: ' "They'll do everything they can to separate us," I said darkly, for I had adopted Chalmers's phraseology and ideas, lock, stock and barrel, and now talked exactly like him: "Every possible bribe will be offered" ' (p. 15). It quickly becomes clear that the caricatured portrait of Chalmers is meant to contribute to the reader's understanding of Isherwood's own fatal fascination for escapist fantasy. For Chalmers Mortmere soon became a state of mind. Wanting to become a poet, he found that he couldn't 'write a single line which isn't *strange*' (p. 74). Isherwood saw himself to be 'a much more complicated psychological mess' (p. 74). One instance of how much further he was removed from the real world than Chalmers occurs when he decides deliberately to fail his Tripos papers by answering them facetiously. Summoned back to explain himself to the college tutor, Isherwood is brought face to face with the hallucinatory nature of the warfare he has been conducting with the dons in his head:

My act now seemed more than ever unreal to me: failing the Tripos had merely been a kind of extension of dream-action on to the plane of reality. How was I to tell the tutor that we had often plotted to blow him sky-high with a bomb? How was I to tell him anything? The tutor wasn't the tutor: he was a kindly but aggrieved middle-aged gentleman with whom I now sat face to face for the first time in my life (pp. 82–3).

Here Isherwood is placing in ironic juxtaposition the literary invention of his over-heated imagination, the caricatured don of the Mortmere stories, and a deliberately underkeyed portrait of the real-life model. It is the same fictional device that he repeatedly employs in, for example, his Berlin novel, *Mr Norris Changes*

Trains (1935), where the narrator's naive and romanticized version of Norris is neatly undercut by Norris's own immoral actions. Isherwood uses caricature as an ironic instrument for uncovering the hidden depths of a character, especially of himself as protagonist.

Chalmers is not the only portrait of an artist apart from Isherwood in *Lions and Shadows*. There is a whole gallery of them, nearly all of whom were destined to become successes in the following decade. But what characterizes them in their late teens and early twenties is a shared neurosis that is most pronounced in the protagonist himself. Deprived of the glamour and danger of active service in the Great War, they all substitute private tests for themselves, mock battles with their parents and their educational authorities. There is Philip Linsley (Hector Wintle), 'chronically unhappy', dying to become 'a brilliantly successful society novelist', hypochondriacal like Isherwood, who 'understood perfectly' Isherwood's 'complex about "War" and "The Test"' (pp. 56–7). Or there is Stephen Savage (Stephen Spender) who 'inhabited a world of self-created and absorbing drama, into which each new acquaintance was immediately conscripted to play a part' (p. 173). According to Spender, this is precisely what Isherwood did, stage-managing meetings between the real-life models for his characters in the Berlin books and generally taking a far more active part in their life than his novels suggest.[41] Like Isherwood, too, Savage suffers from a neurotically induced physical complaint – nosebleeds in Savage's case, throat infections in Isherwood's. The description of Savage's symptom is a good example of the way Isherwood uses caricature indirectly to comment on his own case: 'Without the least warning, at all times of the day, the blood would suddenly squirt from his nostrils, as if impelled by the appalling mental pressure within that scarlet, accusing face; and no keys, no cold water compress could stop it until the neural wound had, as it seemed, bled itself dry' (p. 174). Isherwood belongs to a generation of neurotics and shares with them the physical symptoms of a common neurosis. Elsewhere in the book he makes just as much fun of his own inflammations of the tonsils – what Auden termed 'the liar's quinsey'.

Isherwood constantly catches himself reproducing some of the neurotically induced mannerisms of his friends. On one such occasion in the autobiography he reflects: 'At least seventy-five per cent of my "personality" consisted in bad imitations of my various friends' (p. 147). As one critic, Alan Wilde, has observed, each caricature-portrait in the book 'is in some way an objectification of a quality, good or bad, that is inherent or potential in him'.[42] No character is more caricatured than Isherwood, the protagonist. He presents himself to the reader as his own 'modest exhibit in the vast freak museum of our neurotic generation' (p. 134). Having formulated his myth of 'The Test', that substitute for the experience of real warfare, Isherwood shows how he failed one test after another, selling his motorbike after he 'had ceased to get a neurotic pleasure out of being afraid of it' (p. 59), failing to take sides during the General Strike until it was too late for him to be of use, and feeling seasick when rowing trippers out to sea off the Isle of Wight 'having failed yet another test' (p. 154).

The central metaphor Isherwood employs for his neurosis is his attempt to become a writer. The protagonist outdoes both Chalmers's surrealist cult of strangeness and Linsley's 80,000 words of middle-class gloom in sheer pretentiousness. As Evelyn Waugh wrote in his review of *Lions and Shadows*, 'No biographer of the Strachey school was ever more ruthless in diagnosing affectation, pretension, humbug and secret fear in his subject than is Mr Isherwood in his earlier self.'[43] His portrayal of 'Isherwood the Artist' is a devastating piece of self-caricature in which art and sickness become identified in his earlier self:

Isherwood the artist was an austere ascetic, cut off from the outside world, in voluntary exile, a recluse . . . Now that Isherwood had taken the vow of abstinence from the world . . . it became natural to think of him as being a kind of invalid. Had I ever been seriously ill, I should no doubt have been scared out of my mental bath-chair soon enough; but my health . . . was excellent. So it was quite easy for me to imagine myself subtly and incurably infirm. I remained so, off and on, for the next five years (p. 60).

The modulation from the third person, the object of the protagonist's fantasy, to the mocking commentary of the first-person narrator gives ironic emphasis to the sham pose of the earlier Isherwood. Isherwood the aspiring novelist 'imagined that "being an artist" was a kind of neurotic alternative to being an ordinary human man' (p. 76). Isherwood the author introduces a real artist, Cheuret (André Mangeot), to provide 'a salutary object-lesson to "Isherwood the Artist"':

Cheuret wasn't and didn't in the least feel himself to be 'exiled' from the world. My conception of 'Isherwood the Artist', the lonely, excluded, monastic figure, was something he could never have understood. He never asked for or expected any kind of preferential treatment . . . on the strength of his 'perceptions' or 'temperament' or 'nerves' (pp. 91–2).

As in the case of the college tutor, Cheuret is given a low-keyed, realistic treatment to highlight the absurdity of the caricatured protagonist's unhealthy fantasies.

The first half of the autobiography shows Isherwood heavily under the influence of Chalmers and Linsley. With the entry into his life of the Cheurets his neurotic propensity to revert to childhood fantasies is checked. From this turning point on in the book virtually all the characters contribute to his growth in self-awareness. There is Chalmers's own discovery of E. M. Forster's 'teatabling' of events and his eventual discovery of a formula for transforming private fantasies 'into valid symbols of the ills of society' via Communism (p. 168). There are minor characters like Polly East (Stella Wilkinson) who instinctively perceives the neurosis underlying such habits as his obsessive tidiness and mercilessly points them out to him. Above all there is Hugh Weston (W. H. Auden) who is introduced in the chapter following that in which the Cheurets make their impact on him.

Weston, although the younger of the two, is a portrait of the committed workmanlike artist which Isherwood aspired to be. No-one could be further removed from the stereotype of the diseased and dying artist (*pace* Baudelaire and Barbellion) than Weston with his zest for food and sex, his outrageous behaviour in public and his schoolboy ambition to become a mining engineer. 'People who understood machinery', reflects Isherwood the Artist, 'were doomed illiterates' (p. 114). Instead Weston's early poems surprise Isherwood by their sheer competence. Weston's modern ideas about art are rendered with all Isherwood's comic expertise, as is the heightened picture of his personality which struck Isherwood 'as being, so to speak, several sizes larger than life' (p. 132). On Weston's departure after visiting Isherwood at the seaside in 1926 Isherwood finds that he is suffering 'all the acute mental discomfort of a patient who has been deserted by his psychoanalyst in the middle of the analysis' (p. 121). The use of this particular image shows Isherwood as author anticipating in his choice of simile the role Weston is allotted in the final chapter of the book.

Still a mixture of sense and nonsense, Weston on his return from Berlin is full of ideas of the American psychologist, Homer Lane, whom he has learnt about from a follower, Barnard (John Layard, the anthropologist and psychotherapist). Isherwood as author renders Weston's crude application of Lane's theories to the young Isherwood with farcical gusto. Weston is simultaneously ridiculously priggish (praising the 'pure in heart' because, for instance, they cannot contract syphilis as they are without sexual guilt) and devastatingly accurate in his diagnosis of Isherwood's own neuroses. Isherwood's caricature of the psychoanalyst in the person of a metamorphosed Weston makes the protagonist realize that his art and his life are both suffering from a form of infantile fixation. Even his unrealistic decision to enter medical school, he finally sees, was motivated by 'fear that somehow, somewhere, I should be isolated and trapped, far from the safety of the nursery and Nanny's apron, and compelled to face "The Test" (p. 187). Through the medium of Weston the teachings of Lane and Barnard have become internalized voices in Isherwood's head urging him to break free, 'to shock Mummy and Daddy and Nanny' and 'start growing now' (p. 188). Isherwood concludes: 'As long as I remained a sham my writing would be a sham, too' (p. 187). So he sets out to follow Weston on a decisive journey to Berlin, to establish his independence there and to learn to become a true writer. Unlike his numerous journeys of flight throughout the autobiography, epitomized by his escapist trip north with Bill Scott to Cape Wrath in Scotland, this journey with which Isherwood ends the book is described as part of the journey of life. It is a journey away from the safety of the nursery and the fantasy life of the pseudo-artist which he shared with a younger Chalmers. Ahead lie Weston and Barnard, truth, self-knowledge and the profession of a working novelist. In his capacity as autobiographer, Isherwood applies his dictum concerning the connection between sham writing and sham behaviour by rigorously caricaturing the sham nature of his earlier conduct with all the conscious control of the committed artist.

Isherwood's account of 'the education of a novelist' is simultaneously an account of his changing allegiances, of the friends whom he introjected in the course of a late maturation. As he explains in his note to the reader, because his autobiography 'is about conduct' he has 'had to dramatize it'. It is his dramatic enlargement of his friends which enables him to spotlight those aspects of their characters which became a part of his own conduct and growth. When decades later Isherwood wrote a biography of his parents, *Kathleen and Frank* (1971), he announced that the book 'is really autobiography', mainly 'because it's all explanations about why I am the way I am'.[44] The same could be said about the major portraits of those surrounding Isherwood the protagonist in *Lions and Shadows*. The fact that he has had to borrow from fiction his use of caricature only confirms his conviction that all his novels 'are a kind of fictional biography', just as, up to a point, 'autobiography is always fiction'.[45] The point may be quite low on the fictional scale, as it is in Orwell's case, or self-consciously high as is particularly the case in the writers of comic autobiography that I will be looking at in the next chapter. But Isherwood avoids the confusion between the two genres cultivated by Sillitoe and many modern critics of the genre by insisting that autobiography and fiction are beyond that notional point distinctive ways of interpreting a writer's life and beliefs.

4

The Comic Perspective

I have argued that autobiography as a genre has a considerable degree of overlap with fiction. Historically each genre has influenced the other at different times. But both literary traditions exist within wider generic groupings of which comedy and tragedy are the most generally recognized. In choosing to represent his life in either of these modes an autobiographer is involving himself in more than the adoption of a point of view. He is subscribing to a generic tradition of such age and potency that it is likely to superimpose many of its own norms and procedures on the unwitting autobiographer. I have already cited one instance of this from *My Life and Loves* where Harris is taken over by the tragic mode when his mistress proposes a suicide pact. Because the material he is handling is not the stuff of tragedy his use of a tragic tone exposes a falsity in the writing and undermines the credibility of the episode. Clearly the writer's attitude to his past life has genuinely to coincide with the mode he is employing if the demands of both genres are to be satisfied.

It might be helpful to remember what was understood originally by the term 'comedy'. The word itself derives from the Greek word *komos*, meaning 'revel' or 'merry-making'. According to Aristotle's few remarks on comedy in the *Poetics* the subject matter will tend to be mean, characters will be represented as worse than they are in real life, the outcome will be happy, while the overall aim is, as Molière interpreted it, 'to correct men by amusing them'.[1] Our understanding of the genre has improved considerably since Aristotle's time; but comedy is still getting written that conforms to these criteria. So it should be informative to ask what implications these characteristics have for an autobiographer employing this form of the comic mode. In the first place the central character, his past self, will be the main subject of his merry-making. He and others around him will be represented in unheroic situations and their failings made the subject of the reader's laughter. Comic autobiography would appear to appeal to life's masochists were it not for the presence of the older and wiser author making mock of his earlier life in order to demonstrate his happier state at the time of writing. He has at least corrected himself in the process of giving comic expression to his past existence. A scenario, in other words, pre-exists to which any autobiographer subscribing to this particular view of comedy will have to conform.

The presence of this impersonal element in the most personal of genres has been highlighted in recent times by the Structuralists' demonstration of what they have termed 'intertextuality', the dependence of any one text on others for its meaning. In his autobiography Barthes further complicated this issue by asserting that the self is 'itself already a plurality of other texts'.[2] A simple instance of this was seen to be the case with Conrad whose desire to go to sea was first awoken by his reading of Hugo's *Toilers of the Sea*, Marryat and Cooper. When he comes to write about this in his autobiography the text he produces is further indebted to these and other texts he has read and written. So the part played by intertextuality in autobiography is especially complex. Michael Sprinker has suggested how Vico rather than Rousseau could be claimed to be the father of modern autobiography because his *Autobiography* 'is a text about texts'.[3] Written in the third person, the *Autobiography*, Sprinker argues, 'is not so much an account of himself as it is a narrative of the works that combined to make him "a man of letters" '. His life is represented then as 'a long series of revisions of previously written texts'. Add to this already intricate set of relationships the further set of generic expectations appertaining to comedy and the result in the hands of a writer aware of the potentialities of his chosen mode can be fascinating in its complexity.

There are those autobiographers who abjure the use of comedy for autobiographical purposes on the grounds that, as Graham Greene explains in his autobiography, 'it falsifies history'.[4] Greene, however, is well aware of the element of falsification in all forms of autobiography. His stand represents a personal predisposition to occupy a point lower on the fictional scale than comedy (which he has employed with skill in a number of his novels) permits. On the other hand other autobiographers have chosen to adopt a comic approach to their life stories as a means of countering their inclination to indulge in self-pity or egotism. H. G. Wells takes this point of view to an extreme by arguing that because 'we all have shortcomings . . . any creative writer's life story will be a comedy'.[5] Like Greene, Wells is trying to generalize a personal predilection. After all, the heroes of Greek tragedy have shortcomings which are the cause of their downfall. Moreover Wells appears to be subscribing to the Aristotelian view of comedy in which the stock comic character in Attic drama was invariably the object of invective and ridicule. In *Jokes and their Relation to the Unconscious* Freud voiced a rival view of comedy, one in which the comic protagonist is not ridiculed for his anti-social behaviour but wins approval for it by his display of wit and humour. This idea has had considerable influence on twentieth-century comic writers, perhaps because the comic, according to Freud, 'arises from the uncovering of a mode of thought that is exclusively proper to the unconscious'.[6] It enables an individual to give covert expression to what is otherwise socially unacceptable.

One such unacceptable thought is the Oedipal wish to kill off a parent. As early as 1926 Ludwig Jekel wrote an essay, 'On the Psychology of Comedy', in which he argued that a reversal of the normal Oedipal conflict lay at the heart of comedy. A fellow neo-Freudian, Martin Grotjahn, succinctly explains this theory in *Beyond Laughter* (1957):

The psychodynamics of the comedy can be understood as a kind of reversed Oedipus situation . . . The son plays the role of the victorious father with sexual freedom and achievement, while the father is cast in the role of the frustrated onlooker . . . The clown is the comic figure representing the impotent and ridiculed father . . .[7]

The yield in pleasure for the writer lies in the surreptitious circumvention of those familial taboos that he has induced his readers or audience to break by laughing at them. As Jekel wrote: 'The ego, which has liberated itself from the tyrant, uninhibitedly [vents] its humour, wit, and every sort of comic manifestation in a very ecstasy of freedom.'[8] In the case of autobiography a real, not a fictive, father would have to play this tyrannical role. According to

Freud the father is a major source of the superego. It would appear that a number of autobiographers have cast their fathers in the role of buffoon out of a need to silence the clamorous demands of their over-insistent superego.

John Mortimer: *Clinging to the Wreckage*

A good example of this primary comic scenario occurs in John Mortimer's autobiography, *Clinging to the Wreckage* (1982). The book is almost as much about Mortimer's father, a successful barrister in Probate and Divorce, as it is about himself. The father, we learn in the first section, has become blind, a fact which his son discovered accidentally only on his thirteenth birthday. This is the first of many reversals of the classic legend. In the original story it is of course Oedipus, not his father Laius, who became blind in later life. Yet Mortimer's father did harbour an unconscious wish to inflict his blindness on his son. In the autobiography this is made comically harmless by burying its expression in a welter of inappropriate Shakespearian quotations with which the father bombarded the uncomprehending young boy. The quotation in question is from *King John*: 'He would look at me in a threatening manner and say, casting himself as Hubert and me as the youthful Prince Arthur, about to be blinded: "Heat me these irons hot . . ." '.[9] Hamlet, that classic model of the Oedipal hero, provides his father with another of his favourite Shakespearian imprecations.

Mortimer has said that the main problem in writing an autobiography is 'how to deal with yourself without seeming either intolerably smug or phonily humble'.[10] In his own autobiography he solves this dilemma by taking a comic view of his unsatisfactory past life. As a young boy he was undoubtedly treated ignominiously by his egotistical father. In the book Mortimer distances this upsetting experience by a deft use of humour at his father's expense: 'Busily engaged on his legal practice my father seemed, no doubt understandably, anxious to postpone his complete introduction to me, his only child' (p. 7). Mortimer's skilful use of irony and circumlocution avoids self-pity which could so easily have alienated the reader. Within these comic safeguards he finds the confidence to recount, for example, how on family holidays his father would 'billet' him in some boarding house at the opposite end of the parade to where his parents would be staying in a four-star hotel. More than once he quotes from the diaries his father dictated to his mother to show how unimportant a position he occupied in his father's mind:

Turning the thick volumes, written out in my mother's clear art-school handwriting, I can find out exactly what went wrong with the peas in 1942 and how they coped with greenfly on the roses. It is harder to discover when I was married, had children, got divorced or called to the bar, although most of the facts are there somewhere, stuck at the far end of the herbacious border (p. 47).

Beneath the comic tone lies the realization that he was never more than a lesser bloom in his father's garden. Throughout his life his father had an extraordinarily low opinion of him. He even discouraged him from becoming a conscientious objector during the Second World War, advising him to 'avoid the temptation to do anything heroic' (p. 6). No wonder Mortimer, when he was leading his blind father by the Embankment for a walk, felt a strong Oedipal impulse 'to shake him off, to run away . . . to allow him to wander off, hopelessly among the trams' (p. 5).

Instead Mortimer turns his father into the unheroic butt of his comic autobiography. He shows his clownish old dad insisting on his wife reading out loud the most embarrassing parts of the evidence for his day's divorce case in the crowded commuter train up to town, stumbling into the front stalls of the theatre halfway through the first act and then demanding in stage whispers that his wife explain the parts of the plot he had missed, shouting at an erring solicitor, 'The devil damn thee black, thou cream-faced loon!' or demanding of his wife, 'Are you a complete cretin?' (p. 15). At the mid-point of the book Mortimer allots to his father the part of, not Laius, but Oedipus in his jealous reaction to, not the mother but his prospective daughter-in-law. 'No marriage I could possibly have contracted', Mortimer begins this section, 'could have been more inconvenient from my father's point of view' (p. 93). In opposing the marriage, his father 'chose to proceed, not by pointing out to me the dangers of marrying Penelope, but by persuading her that I was a hopeless proposition' (p. 93). The young couple easily outsmart the father's ludicrous attempts to keep them apart, just as Mortimer ignores his father's advice to abandon writing for the law and finally wins acclaim for his plays even from his ageing parent. As the autobiography draws on, the

father gradually reverses roles with his son, eventually even drafting
the paperwork for his son's court cases. Three-quarters of the way
through the book the father dies after staging a tantrum which he
excuses by explaining: 'I'm always angry when I'm dying.' The
ridiculously inappropriate inclusion of the word 'always' deprives
the father of his last chance to win back the reader's respect. His
dying words mark him as a life-long buffoon. The final section of
the book shows Mortimer, 'the single guardian of a secret we had
once shared' (p. 148), slowly assuming more and more of his
father's role, taking over his chambers, telling his stories, and
restoring his garden to its early splendour. The young comic hero,
once he has defeated the tyrannical parent, has to be seen to be in
possession of the power for which he fought, according to this
version of the genre.

Over a decade earlier Mortimer had given humorous expression
to his relationship with his father in a play, *A Voyage Round My
Father*. The autobiography, which re-uses much of this material,
opens with a reference to the impending theatrical première of this
play in which, he writes, 'a man who had filled so much of my life
seemed to have left me and become someone for other people to
read about and perform' (p. 4). This is precisely the sensation
Borges described in his autobiographical piece, 'Borges and I'. In it
Borges posits a past self, 'Borges', who constantly preempts his
present self, 'I'. As soon as he writes a word it becomes part of that
artificial construct, 'Borges': 'I go on living, so that Borges may
contrive his literature,' he writes. 'I shall subsist in Borges, not in
myself (assuming I am someone) . . .' His final sentence humorously
applies the theory to the autobiographical text in which he has been
expounding it: 'I don't know which one of the two of us is writing
this page.'[11] Both at the beginning and the end of his book
Mortimer compares his advocacy in the celebrated trial of the
editors of *Oz* magazine with his play opening simultaneously at the
Greenwich Theatre: he was awaiting verdicts on both counts. Life
turns itself into a comic play, he infers, into a text, in fact. The
comic writer, he argues, can turn 'anger and misery, defeat,
humiliation and self-disgust' into 'a sense of achievement as he fills
his pages'. 'And yet,' he concludes, 'the catharsis is often too
complete, the life he has led vanishes into his work and leaves him
empty' (p. 197). Having overcome the pain of the past by

converting it into a comic text, the autobiographer finds the text usurping his sense of life. What is signified has assumed an independence from its signifier.

The emphasis Structuralists and their successors give to the primacy of the text is particularly fruitful in the exegesis of comic writing, especially in the case of comic autobiography. The choice of a comic viewpoint shows especially clearly the way in which the writer's life history is turned into a literary artefact which stands at a further remove from the already selective and distorted memories of the past. Mortimer himself obviously understands the consequences of choosing to tell his own story in comic form. As he says, it entails surrendering the self to the text that the self has helped to give life to. 'Suddenly you cease to be real and become a character,'[12] he explained to Sheridan Morley. But comedy also provides Mortimer with an appropriately non-naturalistic form in which to write about events which have ceased to be naturalistic as soon as they entered his memory and inevitably become even less naturalistic in the process of being written about. Comedy, Mortimer asserts, is by definition removed from the illusion of naturalism. He maintains that with its aid 'it is possible to take off and break the sound barrier of pure naturalism' (p. 128), to break through, that is, the illusion that one's past life can be recaptured accurately, by foregrounding the distortions that the autobiographical act entails. But a price must be paid. In turning his father into a comic character, Mortimer concludes, in 'giving him to other people, I came, after a time, to lose him for myself' (p. 197). Once he has written his autobiography, the people and events of his past life have 'left to go into a book'. The text has assumed the life that was his and the author has learnt to see himself as a signifier rather than a creator of the text which he has formed out of his life.

J. R. Ackerley: *My Father and Myself*

Joe Ackerley's posthumously published 'memoir', as he insisted on calling *My Father and Myself* (1968), offers an interesting corrective to the neo-Freudian assumption that all comedy can be reduced to a reversal of the Oedipal scenario. Far from upstaging or replacing his father, Ackerley seeks to come closer to him through understanding him better in the course of writing his memoir of them both.

Freud himself never attempted to restrict comedy to the Oedipal pattern. Northrop Frye offers a much broader definition of what he calls the *mythos* or pregeneric plot of comedy than that employed by neo-Freudian critics. Like Freud, Frye claims that comedy makes for 'an individual release which is also a social reconciliation'.[13] Comedy enables us to transcend the tragic, because comedy, he argues, is 'based on the second half of the great cycle, moving from death to rebirth, decadence to renewal, winter to spring, darkness to a new dawn'.[14] In Frye's view tragedy 'is really implicit or uncompleted comedy'.[15] Frye's position, then, is very different from that of Aristotle.

Ackerley believed that his life had been a failure from beginning to end: 'I have never been happy,' he wrote to Sonia Orwell three years before his death, 'nor ever can be, I was not equipped to be that.'[16] It is all the more intriguing, therefore, that he chose to write his memoir in a gently ironic vein. The self-analysis in which his autobiography involved him he found painful, since it revealed to him a bad failure in communication between himself and his father in particular and the world in general. By resorting to comedy he is able to achieve a form of reconciliation in the book which had eluded him in life: 'It is ever so sad,' he wrote to Donald Windham, 'but rises above itself in a cackle of laughter. That is the way to look at life . . . Ah to laugh, and to laugh at ourselves, if only we can do that.'[17] In *My Father and Myself* Ackerley uses laughter to achieve precisely those effects which Frye identified as being the essence of comedy – individual release and social reconciliation – especially with his dead father. If one thinks of the comic plot in Oedipal terms then Ackerley's is not a reversal of the story of Oedipus but a negation of it.

My Father and Myself has an extraordinary story to tell, as Ackerley was the first to realize. He spent thirty-four years of his life (from 1932 to 1966) perfecting it. Both father and son had unusual personalities. The father, seemingly a respectable business man, revealed to his son after his death (from syphilis) that he had supported a second family (the children of which he had sired). When the son began delving into his father's past he also found evidence that his father in his youth had had a possible homosexual relationship with a wealthy count, and that he had married his mother only twenty-four years after she accidentally became pregnant by him. Not that Joe Ackerley was much more open than his father. He managed to conduct a highly promiscuous homosexual life while sharing the family home for years without ever confessing his sexual proclivities to his parents. As he says in the book, he spent most of his life in a fruitless search for the Ideal Friend which he finally abandoned in favour of an Alsatian bitch on which he lavished all his thwarted affections during its fifteen-year life span and about which he wrote a book, *My Dog Tulip* (1956). Out of this promising material Ackerley fashioned a minor comic masterpiece. In his brief Foreword to the book he explains why he felt it was necessary to confuse the chronology in it: 'The excuse, I fear, is Art.'[18] Art necessitates the redistribution of those shocks which all occurred bunched up toward the end of the story. Ackerley contrives to redistribute the shocks with wit and skill, employing techniques of anticipation, suspense and surprise that one associates with dramatic comedy and farce.

Chapter openings and endings are particularly well manipulated, starting with the first sentence of the book: 'I was born in 1896 and my parents were married in 1919' (p. 11). The conventional beginning of an autobiography is no sooner evoked than it is undermined by the introduction of the unconventional second date. At the end of Chapter 3 he sets out to mystify the reader in the course of describing his father's discharge from the army: 'He brought away two . . . things, the seeds of success in life and the seeds of death' (p. 24). Far from satisfying one's curiosity he only raises it further by postponing the explanation of the seeds of life (his success in business) to Chapter 5 and of the seeds of death (syphilis) to Chapter 9. The transition between Chapters 5 and 6 is a typical example of how Ackerley combines the elements of anticipation and suspense with that of comedy to achieve both a distance from his model and tolerance for his father's philanderings. Chapter 5 ends:

The year 1892, therefore, was a momentous one in my father's life; in it he lost a wife, gained a job which was to make his name and fortune, and picked up my mother on a Channel boat. She was not, as we have seen, his first conquest, nor was she to be his last, and it may therefore be permissible to wonder how long the affair would have continued had not an accident occurred (p. 48).

Chapter 6 begins: 'My elder brother Peter was the accident. "Your father happened to have run out of french letters that day," remarked my Aunt Bunny with her Saloon Bar laugh . . .' (p. 49). Ackerley has already mentioned his mother's accidental pregnancy in Chapter 2, but his comic manipulation of this not inherently funny material places the father's subsequent potentially blameable actions (like hiding his wife and children away in Cheshire for nine years) in an understanding and more forgivable light.

The comic perspective, then, as Frye has suggested, enables an autobiographical writer, such as Ackerley, to transcend the tragic pattern of his life if only retrospectively in the course of constructing his comic text. The text is based strictly on the small amount of hard evidence Ackerley could assemble, yet its comic expression gives it an independence from his own life which he consistently characterized as a failure: 'I am just a bad character, that is all, cruel and heartless. I don't care for people . . .'[19] he wrote in his diary. In one sense the memoir is a contradiction of his assertion. In another sense it is the product of the genre which Ackerley chose to adopt, a genre which forced him to see his life and so describe it in an uncharacteristic light. At the end of the book he recalls keeping a diary for six months in the hope that from it 'a book would emerge more fascinating, however clumsily written, than if he had been anthropologizing among the Pygmies' (pp. 215–6). But on rereading the diary fifteen years later he found that it 'contained no single gleam of pleasure or happiness, no philosophy, not even a joke.' Instead it was 'a story of unrelieved gloom and despondency, of deadly monotony' (p. 216). A diary tends to reflect more of the personality and less of the reflecting artist responding to the generic requirements of the form. So disparate are these in Ackerley's case that it is little wonder he spent thirty-four years perfecting his mildly comic treatment of potentially tragic material.

The gently ironic tone of comedy that Ackerley employs enables him to win greater tolerance from his readers than he does, for instance, in his published diaries. To understand why one has only to compare his slightly mawkish evocation of his mother in his diary – 'childish, gay, sentimental, romantic', 'the sweet creature',[20] with the comically more distanced portrait of her in his memoir – the pick-up who became pregnant, for instance. In the book a very different personality emerges, one, for example, who is unable to cope with his father's terminal illness even at a distance:

Nervous of life, frightened of death, surrounded by sedatives and boxes of glycerine suppositories, she was already forming that eccentricity of habit which was soon to confine her, an affable chatterbox and for a time a secret drinker . . . within the sheltering walls of her own dwellings for another seventeen years. (p. 151).

The distance that comedy requires enables Ackerley to adopt the tone of a mildly ironic biographer to whom the subject is an interesting case of mild neurosis.

Ackerley's father leaps to life on the page as the true comic hero of this autobiography. He is simultaneously offered as a butt for the reader's humour and exonerated from moral censure by Ackerley's subtle use of the comic mode. For instance the opening sentence of the book in which Ackerley's birth in 1896 is juxtaposed to his parents' marriage in 1919 is followed by a sentence intended to pass off the discrepancy with a smile: 'Nearly a quarter of a century may seem rather procrastinatory for making up one's mind, but I expect that the longer such rites are postponed the less indispensable they appear and that, as the years rolled by, my parents gradually forgot the anomaly of their situation' (p. 11). In fact, as he tells us elsewhere in the memoir, his father was not in two minds about the matter. He had no intention of getting married if he could help it. Ackerley's ironic use of understatement here simultaneously turns a moral failing into a mere foible on his father's part. The comic text works towards a form of reconciliation between father and son. His father was evidently a great womanizer; but Ackerley seeks to pre-empt moral condemnation by making fun of his father's weakness while associating it with another form of appetite more generally acceptable: 'In male company he was liable to refer to pleasing specimens of the female sex who caught his eye in the street as "plump little partridges".' When this 'predatory, gastronomic approach to women' displeases the aunt, Ackerley proceeds to air the suspicion that 'she, and my mother too for that matter, never found the sexual act agreeable or hygienic' (p. 20). Ackerley's deft comic touch is epitomized by that inapposite final adjective. One human foible is negated by another and the reader is left free of moral inhibitions to enjoy the humour of the situation. Comic tone also enables Ackerley to reconcile himself, I feel, to some of the more unpalatable facts in his story. His father's early death, for instance, which he so much regretted, is given comic distance by a playful pun which rests on a knowledge that his father was planning to take his second 'wife' to Bad Gastein when he entered his terminal illness caused by syphilis: 'But whatever he understood about his condition, the gravity of it must have been kept from him, for at the time when the doctors knew that he would shortly be dead he had tickets for another sort of journey in his pocket' (p. 91). That is how Chapter 9 ends, punning on an Americanism in a kind of comic inversion of Othello's 'journey's end'.

This exculpatory use of comedy is applied equally to his portrait of himself. Ackerley, as we saw, considered that his life had been a failure, especially his life as a homosexual. As a writer he knew that his homosexuality was liable to prejudice many of the readers of his autobiography. One of his defences is to ask whether he is any worse than his father whose sexual promiscuity he has already attempted to show in a favourable light. Reproachful that when he arrived at his preparatory school he was wholly ignorant of sexual matters, he gives comic voice to his grievance: 'Indeed, considering what I afterwards learnt of my father's behaviour, and of the licence and impropriety of his relationship with my mother, I think it a trifle dishonest of them to have excluded me so completely from that freedom of thought in which they themselves seem to have indulged' (p. 79). The combination of dramatic exaggeration ('licence and impropriety'), understatement ('a trifle dishonest') and circumlocution effects a comic reconciliation between vice and mere thoughtlessness.

Ackerley discusses his homosexuality with unusual frankness in the second half of the book. His witty treatment of his sad experiences with his own sex bears out Freud's assertion that 'the spheres of sexuality and obscenity offer the amplest occasions for obtaining comic pleasure'.[21] Comedy provides a socially acceptable form of expression for the asocial sexual libido. Conventional moral judgements are held in suspense by someone who encapsulates his unsatisfactory love life in so dismissive a sentence as: 'It may be said to have begun with a golliwog and ended with an Alsatian bitch; in between there passed several hundred young men, mostly of the lower orders and often clad in uniforms of one sort or another' (p. 110). The golliwog he asked his astonished father for at the age of 12. The guardsmen whom he picked up in such incredible numbers may have been connected at an unconscious level with his father's youthful experiences as a guardsman. This connection could help to explain his ironically self-mocking assertion that, although 'two or three hundred young men were to pass through my hands in the course of years, I did not consider myself promiscuous but monogamous, it was all a run of bad luck, and I became ever more serious over this as time went on' (pp. 124–5). Just withholding the full stop after 'monogamous' humorously reproduces the breathlessness with which he rushed on to the next lover. Was the Ideal Friend for whom he spent a lifetime searching in vain a reincarnation of his father despite the latter's predominantly heterosexual leanings? 'I never suffered much from jealousy,' Ackerley writes, 'and the Ideal Friend could have a girl friend or a wife if he wished, so long as she did not interfere with me. No wife ever failed to interfere with me' (p. 137). As part of the subtext it cannot be decisively affirmed, but the context that the book provides does suggest such as association.

At various points in the book Ackerley shows that he is aware of the potential conflict between truth-telling and the needs of a comic text. One such point occurs when he is describing his still virgin state at university where he had a platonic affair (for want of courage) with a fellow undergraduate. 'Unable, it seemed, to reach sex through love, I started upon a long quest in pursuit of love through sex. Having put that neat sentence down I stare at it. Is it true?' (p. 123). The use of verbal humour is a constant threat to the autobiographical pact. Yet at the same time the adoption of a comic tone assists self-discovery. In describing his early attempts to turn writer while living at home on an allowance from his indulgent father (who had previously planned a successful career in the law for his son), Ackerley pillories himself without pity: 'I was now set to be a writer, and my Father no doubt made an easy displacement of the Lord Chancellor for the Poet Laureate. My study was understood to be private ground where the great mind could meditate undisturbed' (pp. 102–3). Comic magnification ensures in advance that the reader will not take his youthful dedication to the world of letters over-seriously. Sure enough Ackerley next reveals that he couldn't write anything of value at this time. His guilt consequently made him feel more and more persecuted, though, he adds, 'The only person who persecuted me was myself' (p. 103). Like Isherwood in *Lions and Shadows*, Ackerley is prepared to portray himself with more satirical bite than he allows himself to use on those around him.

Comic perspective also enables Ackerley to view life as a whole in a less morbid and self-absorbed light than emerges from his diaries, because it provides him with that reflective dimension which lies at the heart of all autobiography. The book ends with the application of this perspective to the text itself. After his mother's death, he recalls, he had to go through her personal effects including an array of trunks, suitcases and cardboard boxes. As he expectantly opened each he was repeatedly confronted with nothing but wastepaper: 'That was my mother's comment on life. It might serve also as a comment on this family memoir, which belongs, I am inclined to think, to her luggage' (p. 208). Here the clown turns not on himself but on his clowning. If life is, as the book suggests, a grotesque joke, then this memoir of life must itself be part of the joke, his useless luggage for revisiting his ridiculous life. In a sense Ackerley here is effecting his own form of reconciliation between the potentially conflicting requirements of the two genres. Once his comic text has been written and become part of his grotesque past it becomes another act of clowning and is legitimately another object of the irony he has so skilfully turned on his past self.

V. S. Pritchett: *A Cab at the Door*

According to Frye there are six basic types of comic structure ranging from the ironic at one extreme to the romantic at the other. The most ironic type is that in which the older generation triumphs over the younger or remains undefeated. The second type, Frye writes, 'is a comedy in which the hero does not transform a humorous society but simply escapes or runs away from it, leaving its structure as it was before'.[22] This appears to be the category into which both Ackerley's memoir and Sir V. S. Pritchett's *A Cab at the Door* most naturally fall. Pritchett wrote two volumes of autobiography, *A Cab at the Door* (1968) covering the first twenty years of his life from 1900 to 1920, and *Midnight Oil* (1971) mainly covering his early adult years from 1920 to 1939 during which period he established himself as a professional writer. The first volume is written in the comic mode; the second alternates between comedy and a more serious tone. It is interesting that in the second volume it is the episodes involving his parents which are treated comically as if their presence were, as Frye suggests, an essential ingredient of the comic genre. In *A Cab at the Door*, on which I will concentrate here, the father shares the lead role with his son. True to Frye's prescription, the volume ends with the twenty-year-old Pritchett's escape from his domineering father's control by leaving for Paris.

In a presidential address Pritchett gave to the English Association on the subject of autobiography he described the act of writing about one's life as a choice of genres: 'What we have to decide is what play we are putting on, what its *theme* is and what postures fit into it.'[23] His first volume of autobiography is superior to the second just because in the former he has made a firm decision about what play he is putting on, which is a comedy. The comic genre allows him, as Frye suggests, to throw his main emphasis on the 'blocking characters', his parents, especially his father, and to cast himself in the role of what Frye calls the 'technical hero' who tends to be overshadowed by the larger-than-life comic heroes of the older generation. According to Frye the chief characteristic of the comic hero is that he is 'obsessed by his humour'.[24] In this role he can be seen either as a comic butt or as a socially subversive hero. It all depends on the point of view. If, as Robert Torrance has argued in *The Comic Hero*,[25] you take the Aristotelian view of comedy, which largely reflected the social attitudes of the ruling aristocratic classes, then you will present your comic characters as objects for ridicule. If, on the other hand, you adopt the popular view of comedy which reflects the responses of the less privileged classes, then the comic opponent of the establishment becomes the unconquerable underdog, their heroic spokesman in comic guise.

Pritchett's portrait of his father partakes of both traditions. As a figure of paternal authority which he exercises arbitrarily over his entire family, he plays the role of the *senex iratus* in classical comedy, the overbearing butt of the comic writer's humour. Yet he also belongs to the modern tradition of the invincible subordinate, the Schweik or Chaplin of our day. No sooner is he bankrupted than he appears on the scene more resplendent than ever spending money in all directions. Pritchett's technique for describing these irrepressible urges towards self-indulgence on his father's part is to evoke the child's natural hero worship of his father in order to create an ironic disparity between the child's and the adult's view of the father. So it is the father and not his children who gets to see *Peter Pan*: 'We, my mother, brother and I, are not envious or jealous. We have no desire to see things like the pantomime or *Peter Pan*: other children see such shows, but we prefer to send father there on our behalf; it will be one more chapter of his fantastic life'.[26] The application of the present tense to his past lures the reader into temporary acceptance of the child's naive response. But that final clause looking forward to the time of writing the autobiography alerts one to the fact that Pritchett is not simply reproducing the child's point of view unedited. The amused author has refined it and given it retrospective significance. The same dual perspective is present in Pritchett's description of his father's rise to the lofty position of furnishing purchaser for the aircraft firm for which he worked during the Second World War. Starting as a letter-stamper in the Registry, he attained the dizzy heights of purchaser by cultivating 'a look of owning the place'. The 'child's' over-simple explanation of his father's success, reproducing his father's words as it does, subtly reveals the essential childishness of the father. His father claimed that his earlier life as a salesman 'had taught him to start life from the top' (p. 182). For the protagonist his father makes the impossible seem possible; but the narrator's use of paradox simultaneously undermines the father's seeming accomplishment. With a puff of smoke and pink flame his father, the magician, transforms himself into the creator of scores of offices and square miles' worth of carpeting and linoleum. He is constantly being dined out by furniture salesmen: 'He brought back menus from Claridges and the Ritz and went over the dishes for our benefit. We were proud to have him eat on our behalf and admired the new sad look of the gourmet under his eyes' (p. 182). Where the child is content to let his father live his son's life by proxy, the adult Pritchett humorously indicates the father's abrogation of his unsuspecting son's needs. The ambiguity produced by this dual perspective is due to the double role that the father plays in the book. He might function as the irrepressible comic hero but he is also a tyrannical figure of authority who can turn on his dying niece for daring to suggest that she is ill or who can rant at V. S. Pritchett's brother for daring to take himself off to France where he learnt the silk trade after his father had refused to employ him in his own firm. On his return from France he is lambasted by his father for daring to travel back with his French manager:

Father continued his jealousy: on the Monday my brother would be sacked for his presumption.

'I cannot understand your lack of judgement,' he said, 'You go to the same lavatory and then – it's past belief – you travel in the same first-class compartment with him.'

'He asked me to. He doesn't speak English.'

'What! You little fool, every Frenchman speaks English!'

On Monday morning my backward brother was promoted (p. 205).

The narrator's insertion of the father's use of 'backward' without quotation marks neatly turns the accusation back against its originator without overt intervention by the adult Pritchett. Here is the figure of the *senex iratus* receiving his come-uppance from the younger generation. But he continues to fight his wife and sons till his death in his eighties. He may be a 'romantic procrastinator' and 'dreamer', but he is also 'a pocket Napoleon' who considers himself beyond criticism (pp. 206–7).

Pritchett makes full use of his father's conversion to Christian Science to illustrate with humour this unique combination of oppression and ebullience. As a child Pritchett was struck by the image of a poster for a play called *The Bad Girl of the Family* in which an evil-looking man stands in the doorway of a bedroom where a woman in evening dress sits doing her hair. Pritchett as autobiographer mischievously confuses this villainess with both Miss H, his father's business partner and sometime mistress, and Mary Baker Eddy, the founder of Christian Science. So the child explains to himself his father's abscondment with Miss H at one point by deciding that he had gone off to be with the prophetess of Christian Science. The child's view of her as the Bad Girl of the poster allows the adult Pritchett to treat her as a villainess in the life of the family. What he says of the poster is true of his autobiography: 'It was a mixture of our life and fiction' (p. 69). But the adult autobiographer's use of this piece of fiction enables him to voice his genuinely felt criticism of Christian Science. Mary Baker Eddy did fulfil in a number of ways the role of villainess. Her teachings encouraged his father constantly to evoke the Divine Mind to justify his arbitrary decisions, especially those concerning his own family. How could the young protagonist contradict a

father who was merely 'voicing' an unwelcome 'Truth'? Pritchett
makes hilarious fun of the terminology originating in Mary Baker
Eddy's books which his father adopted with such enthusiasm. On
the one hand the father comes over as one of life's irrepressible
victims who refuses to go under by falling back on the comforting
doctrines of Christian Science, letting God's will 'unfold' when
sitting on his backside at the office surrounded by unpaid bills and
persuading himself that 'behind all the human sources of money,
there was the great investment house of God pouring out wealth'
(p. 206). Here is the invincible underling of the popular tradition
refusing to allow mere matters of fact to interfere with his
grandiose dreams. But, as Pritchett points out, 'righteousness killed
his heart'. With the departure of his former gaiety the angry tyrant
took over. 'Self-punished, he slowly drifted into punishing us'
(p. 83).

Some of the supporting cast come straight from low comedy or
farce. Even his mother is treated more in this fashion. So afraid of
callers that she denies her identity to enquirers at the front door, a
miracle of ineptitude in home dress-making, she is portrayed as the
comic butt of her husband's impractical fantasies. In her turn she
takes her revenge on him by attacking, not him, but everything to
which he is attached. His business partner becomes 'that woman';
his religion becomes the excuse for ruined Sunday lunches, while
his expensive suits take more material punishment: 'She had a
vengeful streak in her, and looking at our father, the impressive
Managing Director, and counting his suits and knowing she
couldn't get a penny out of him for our clothes, she attacked his
wardrobe' (p. 90). As a result the young Pritchett lands up at school
with her cut-down version of the Managing Director's trousers the
flies of which quickly open up at the seam to leave him indecently
exposed by the end of the first day's wear. Repeatedly he depicts
himself as the comic victim of his parents' running battle with each
other, parents who are both also victims, although they have rather
more control over their predicament than the child did. Some of the
minor characters are made to conform to the more farcical
possibilities of the comic genre. There is Great Uncle Arthur, a
carpenter who seems to be continually chewing nails with his teeth,
married to genial Aunt Sarah with her 'large, loose, laughing teeth
like a horse's', the two of them doting on one another: 'It seemed

that two extraordinary sets of teeth had fallen in love with each other' (p. 41). Or there is Aunt Lax, married three times, increasing her fortune on each occasion: 'These marriages were a shock to the family, but Lax in name, Lax in nature . . .' (p. 48). In both these instances the verbal wit that excludes the possibility of psychological complexity comes not from the child but from the narrator bent on demonstrating the widespread normality of the farce which was played out with such intensity in the family home.

By contrast, when it comes to his own portrait Pritchett draws on the double vision of the child and the reflecting adult autobiographer to achieve considerable subtlety within a comic framework. Several incidents are ostensibly told from the child's limited viewpoint while depending on an adult level of understanding for their humorous impact. The young Pritchett's misunderstanding of what adults mean by 'rudeness' or 'Calvary' (which he mispronounces 'cavalry') is used to poke fun at corrupt adult values. The arrival of another baby, his youngest brother, is seen as 'another betrayal' by the young boy: 'It cannot be a good thing because Mother is always ill and talks about funerals and the illnesses of all her relations, and some woman comes in and cooks mutton stew which we refuse to eat' (p. 57). Although the young Pritchett cannot see it, the older Pritchett uses the innocence of his youthful protagonist ironically to expose the ridiculous postures that the adult world adopts towards that most natural of events, birth. As here things usually happen *to* Pritchett as a child. In this respect too he conforms to the role of the passive 'technical hero' of comedy whose real life, as Frye suggests, begins at the end of the play.[27]

When a writer looks at his younger self in a comic light it seems more likely than not that he will see himself as an innocent victim of circumstances. There is, for instance, the incident of Pritchett's near-castration after sitting down on glass: ' "Another inch", says the doctor, "and he would have had them off." What? I get a fort, soldiers, a fire in my bedroom for this. It is worth it' (p. 55). Again it is the child's viewpoint that is responsible for the humour. 'I notice', Pritchett wrote in 1948, 'that I am less of a Hotspur and more of a Falstaff every time I get a challenge.'[28] Twenty years later in his autobiography he has virtually eliminated the Hotspur in him that would seek to do battle with the world, and especially with his father. The only suggestion of a core of Oedipal resentment emerges indirectly when Pritchett describes a bullying old coster and his cowed son with whom Pritchett does occasional business once he has gone to work in the leather trade:

On one of his visits, the son went on about Anarchism.
  'If he had the orders,' he said to me, 'I say *if* he had the orders, a true Anarchist would even kill his father and his mother. You've got to follow the logic of it.'
  'Would *you*?'
  'Wait a minute – I said "*If he gets the orders*." If they come, well – tit, tit, tit, you've got to admit the logic.'
  The pair seemed to parody my life (p. 180).

Parody is as close as Pritchett comes to voicing feelings more appropriate to a tragic outlook within his chosen comic framework.

Beneath the humorous surface lies what Pritchett has elsewhere called an obsession which he shared with his brothers and sister: 'the obsession', he explained, 'with our extraordinary father and our engaging mother about whose rights and wrongs we cannot stop arguing even today, as if they were the only instances of fatherhood and motherhood on earth. We felt ourselves to be mythical and exceptional: isolated.' One of the characteristics of comedy is to expose the private and the secretive to public view, and in so doing to rid it of its potential threat. 'When I had finished the [autobiographies]', Pritchett recalled, 'I saw that I had rid myself of the mutilating part of my obsessions.'[29] Near the end of *A Cab at the Door* he does offer his own (somewhat sexist) explanation for his father's extraordinary conduct. Departing from his habitual comic tone, he reflects: 'He was one of those fathers who are really mothers; he had a mother's primitive, possessive and jealous love and indeed, behaved as if he and not our mother – who was not possessive at all – had given birth to us. He wished to preserve us for himself' (p. 207).

It is interesting to compare *A Cab at the Door* to an earlier fictional portrait of his father in his novel, *Mr Beluncle* (1951). In the novel he also attempts to portray his father in a comic light; yet it fails to make the same impact. It seems as if at this earlier stage of his life his obsession was still too deep-seated for him to want to stay close to the historical 'truth' as he remembered it. Instead he turned his parents into grotesque figures of fun. The title, *A Cab at the Door*, comes from the constant moonlight flits which the father's business fiascos forced on the family at regular intervals. In the autobiography they are humorously alluded to as those 'journeys which my father thoughtfully provided' (p. 29). But in the novel Pritchett evades the residual pain that the memories of these continual removals held for him by nudging the material into the realm of sheer fantasy. Mrs Beluncle (modelled on his mother) is explaining their tendency always to be on the move to a neighbour:

'It's because we can't *breathe*,' said Mrs Beluncle. 'One day you'll look out of the window and you'll say, "Where are the Beluncles?" Gone. Yes, we may be moving to Balmoral Castle. I expect the King would move out. Or we may just be floating in the air. Or we may just stay. But even if we stay, we'll be moving in our heads, as you might say.'[30]

This escapist element sabotages the novel. But in the case of *A Cab at the Door*, as Graham Hough wrote in his review of it, the autobiographer 'abandons the safeguards of fiction while still exposing himself to the rigours of art'.[31]

Certainly the rigours of comic autobiography involve the writer in satisfying generic requirements quite as demanding as those of the novel. One can further argue that *A Cab at the Door* is a product of intertextuality, heavily indebted to earlier novels – and not just *Mr Beluncle* – for its success. The father is more than once called a 'Mr Micawber', alerting one to Dickens's influence on Pritchett's comic depiction of character. Then there is Uncle Frank who has 'the waxen look of young Chatterton', or the Yorkshire relations who are called tamer versions of Charlotte Brontë's Haworth country folk. But intertextuality is more complicated than this in the case of any autobiography that sets out to describe how the writer took to writing as a profession. The books that influenced him as a child are partly responsible for the book in which they are mentioned. Dickens, for instance, is quickly discarded in favour of comics: 'To hell with poor self-pitying fellows like Oliver Twist; here were the cheerful rich. I craved for Greyfriars, that absurd Public School, as I craved for pudding' (p. 98). Soon he and his schoolmate are imitating the fictitious schoolboys of Greyfriars, shouting words like 'Garoo' and strolling arm in arm round, if not the quad, then the local block. Here the comics he has been reading momentarily dictate not just the subject but the style of the passage in which they feature.

The wider question of the relationship between life and literature is pursued further in the autobiography. His first attempt to write fiction ('Last night I went to see the berth of a child' (p. 59)) causes ructions at home when his father reproves him for telling an untruth. Next he succeeds in writing the best account of a Zeppelin raid in his class by discovering 'the first duty of a novelist – to become someone else' (p. 138). He pretends to be his mother. This audacious lie, he writes, prevents him from indulging his worst tendencies – muddle-headedness and showing off. Is this the origin of the use to which Pritchett puts his parents and others in the autobiography? He seems still to be drawing on this early lesson in literary craft whenever he observes his younger self through the eyes of others. Even when he observes himself through his own eyes, he frequently reproduces the vision and speech pattern of himself as a child who has become, especially in the autobiographical act, someone else.

Books also provide Pritchett with the necessary literary judgement to see through Mary Baker Eddy's awful prose ('Meekly our Master Met the Mockery'). This in turn gives him the confidence to reject her religion which his father had temporarily persuaded him to adopt. It also enables him as narrator to suggest the childishness of his father's ready acceptance of her doctrines. Books were one of Pritchett's main ways of escaping from the confines of his life at home. He claims that he learnt to write by reading J. M. Barrie's *When a Man's Single* and cites H. G. Wells's similar claim in his autobiography, *Experiment in Autobiography* – another convoluted instance of intertextuality. He also attributes his desire to travel to his reading of Belloc, Stevenson and others. His life, which has been greatly formed by books, now takes the form of another book, one which, as Pritchett writes, 'begins a life of its own after it leaves (a writer's) hands'.[32] In fact it began a life of its own from the moment he chose to tell his own story as a comic autobiography.

Frye argues that comedy traces a movement from illusion to reality, from stasis to its opposite.[33] Pritchett's autobiographical comedy conforms to Frye's definition in that it traces a movement from the illusion that his father embodies the truth to the reality that fiction governs all their lives. But to live out that reality he has to go abroad and leave behind him the stasis of his father's presumption of power over his children. At the end of the first volume he leaves the stage to start living out his own fiction: 'I became a foreigner. For myself that is what a writer is—a man living on the other side of a frontier' (p. 211). True to Frye's prescription Pritchett's protagonist or technical hero hasn't changed the society of his childhood but simply escaped from it into the less comical world he describes in the second volume of his autobiography where he puts on a different play in which he casts himself in a different part.

In Search of Identity

5

Childhood

In focusing on problems concerned with autobiographical truth I have tended to concentrate during the preceding four chapters on two of the three constituents of autobiography – *bios* and *graphos*, the life history and the act of writing it. Yet most readers of autobiography look to it above all for the depiction of, or at least the search for, *autos*, the self. Autobiography is potentially the most subjective, because the most self-regarding of genres. The true autobiographer is absorbed by his own nature, its genesis, its evolution, its conflicts. Implicitly he assumes that subjective experience underlies and conditions objective observation. Autobiographers who seek to give precedence to the objective world inevitably produce memoirs or reminiscences in which the writer hides behind the role of fortunate observer. Such books can be fascinating in their own right, but they are generically different from subjective autobiography. One reason why the two genres have become confused with one another has to do with the relative modernity of the focus which subjective autobiography directs on to the individual self. Prior to the Renaissance there was little encouragement to explore one's personal identity as an end in itself. In earlier times self-portraiture is oblique and is usually concerned with the public persona, not the private life of the writer.

Georg Misch, Karl Weintraub and George Landow all argue for a historical conception of subjective autobiography. 'In a certain sense', Misch writes, 'the history of autobiography is the history of human self-awareness.'[1] It could be argued that it took the break-up of the Roman Empire for Augustine to acquire the sense of selfness which distinguishes his autobiography from earlier accounts of personalities conceived as static and representative of the stable society to which they belonged. A similar upheaval during the Renaissance provided the impetus for the rise of modern autobiography with its concentration on the private self. A unique feature of autobiography is the ostensible identity of author, narrator and protagonist. This enables the author to perform a kind of interior excavation, using all the sophisticated literary techniques of a narrator, in search of a protagonist who was, but no longer necessarily is, the same person.

But what is meant by this 'self-awareness' which it is claimed has increased in the course of man's history? The concept is complex and gives rise to a multiplicity of problems. For a start the process of searching for the self can itself become a realization of the self. Such an autobiography transforms its original excavatory intention into one of self-creation. A recognition of the irrevocable pastness of the past frequently leads autobiographers to shift their attention to the present moment of written recollection. Extreme examples of this attitude are found in Barthes for whom the 'I' is 'always new',[2] and in Nabokov who dissolves the identity of his protagonist from the past into that of the narrator communing in the present with his memories. Other problems involve retrospective distortion (whereby, the literary autobiographer predates his sense of vocation in his search for a sense of identity), and the impossibility of seeing the self simultaneously as 'I' and 'he', first and third person protagonist-cum-narrator. All these problems originate from the same source – the desire of the self to know itself from within and yet as if from without. Subjectivity feels compelled to search for an illusion of objectivity that might render the subjective comprehensible to itself and others.

This paradox is central to the nature of autobiography. Autobiography knowingly attempts the impossible. It does this by means of literary indirection, what James Olney has called the use of 'metaphors of the self'.[3] This resort to fictional similitudes bears some resemblance to analysts' use of dreams and neurotic symptoms to construct a picture of the otherwise inaccessible self. It is my opinion that of all the explanations of the phenomena of individual growth and development psychoanalytic theories are the most comprehensive. Psychoanalytic concepts like the timelessness of unconscious memory, the effects of infantile sexuality on later behaviour, or the defence mechanisms of the ego (such as regression, fixation and sublimation) are continually finding authentication in the evidence offered by autobiographies normally written with little knowledge of these concepts. This is why I will continue to refer the reader to leading psychoanalytical explanations of the origins and growth of identity throughout much of this second section of the book. Autobiographical introspection has many resemblances to the analytical process. Both can be mutually illuminating where the primary interest lies in searching for a better understanding of the self.

If the child is father to the man it can be argued that the human life cycle consists of a series of attempts to recover the security of the infant by various substitute procedures. One of these is to relive one's childhood either in fantasy form or in the course of artistic creativity. In 'Creative Writers and Day-Dreaming' Freud argues that the child as he grows up substitutes daytime fantasies for his earlier games, both being forms of wishfulfilment. The only difference between the two activities, he maintains, is that the adult is ashamed of his childish fantasies and hides them from others. The creative writer, on the other hand, shares his fantasies with his readers. At the same time he seeks to win their approval, partly by disguising his portrait of what Freud calls 'His Majesty the Ego, the hero alike of every day-dream and every story',[4] and partly by offering them a yield of pleasure in the form of the aesthetic shape which he gives to his material. As an explanation of fiction this is over-restricted; but his theory has more immediate implications for autobiography, especially for that most fictional of subspecies, autobiography of childhood.

Memories of childhood are most subject to unconscious censorship and alteration. The anxieties that all infants are said to experience give rise to a complicated succession of psychosocial defence mechanisms aimed at protecting the ego from threats to its pre-eminence. Consequently the adult's view of childhood is invariably distorted in favour of the child's successes in dealing with threats to its needs and desires. One critic of Victorian autobiography has called childhood 'the invention of adults' because it 'reflects adult needs and adult fears as much as it signifies the absence of adulthood'.[5] Childhood autobiography, therefore, offers the writer the opportunity of winning public approval for his adult version of his early years. It offers him one more defence mechanism for combating the continuing threats to his ego. Again and again twentieth-century autobiographers recall a childhood spent in a Garden of Eden, what Chesterton calls 'a lost experience in the land of the living'.[6] This process of idealizing the past often receives additional impetus from the way in which adults recount stories about a child's earliest forgotten years to him in later life. Such stories have a natural tendency to enhance the self-esteem of both adults and child. They are 'family tales', as Leonard Woolf observes, 'told so often about one that eventually one has the illusion of remembering them'.[7] Seen in this light, childhood autobiography can represent a further stage in a lifetime's formation of a myth or fantasy about the past that enables the present self to cope with the future.

Despite this inbuilt distortion, most truth-seeking autobiographies hope to achieve greater knowledge of their present selves from an examination of their early formative years rather in the manner of classical psychoanalysis. Edwin Muir's autobiography is an instance in point. He shares the commonly experienced sense of childhood as a garden of Eden from which he was cruelly expelled – in his case at the age of fourteen. He was uprooted from the Orkneys where he had spent an idyllic childhood on his father's farms, and underwent a traumatic five years in Glasgow during his late teens. His neurotic response was to dismiss the past from his consciousness. Then in his twenty-first year he had a crucial meeting with a humanist intellectual which completely changed his attitude to his childhood. The humanist was 'fascinated by the beginnings and development of things', Muir writes, 'a world which, as my eyes were fixed on ends, I had impatiently ignored'. But childhood cannot be that easily dispensed with. 'Gradually I began to realize,' he records, 'as I listened to him, that in dismissing the past I was dismissing all knowledge and all life as life is known.'[8] At first Muir takes 'knowledge' and 'life' to mean ideas and books. Only much later does he come to realize, partly with the help of psychoanalysis, that knowledge is dependent on self-knowledge and life on the life within himself. One of his insights at the end of the book is a realization of how much we receive from the past, a past 'on which we draw with every breath', but also a past which he associates with 'the Source of the mystery itself'.[9] Investigation into his origins put Muir in touch with a fundamental need within him, one which he had kept concealed from himself up to that point.

The conflict between idealizing one's past and seeking within it for greater self-knowledge, between defending the ego and stripping it of its defences in order to understand the origins of adult behaviour, underlies all good childhood autobiographies. It is a conflict that cannot necessarily be any more resolved in autobiography than in life. It is striking how often autobiographers employ paradox, the linguistic equivalent of such conflict, to mirror this sense of ambivalence. The myth of the Garden and the Fall is itself highly ambivalent. Maybe this is why it has attracted such a wide range of childhood autobiographers – because it offers a paradigm for their own complex experience. In many cases it serves the function of converting a very private experience into a commonly shared one. As a primal myth it connects an autobiographer's personal memories of his particular childhood to racial memory. Above all it offers the writer a means of overcoming a difficulty inherent to the genre – how to interest the reader in the story of his life – by making his story parallel the story of all mankind.

Herbert Read: *The Innocent Eye*

Herbert Read's earliest volume of autobiography, *The Innocent Eye*, was first published in 1933 when he was forty. It covers the first ten years of his life which were spent on his father's farm, Muscoates, in the Vale of York, at that time a remote valley in the North Riding. It was not his first attempt at prose autobiography. Previously he had kept a War Diary between 1915 and 1918 and had published two short accounts of incidents at the Front in which he participated, *In Retreat* (1925) and 'The Raid' (1927). 'The Raid' is itself a prose reworking of an early poem, 'Childhood', written during the war. In 1940 he carried his life history forward from his father's death in 1903 to 1939 in *Annals of Innocence and Experience* which in its reissued form of 1946 included *The Innocent Eye* and his two short war narratives. Finally in 1963 he added the War Diary and 'A Dearth of Flowers', a series of meditations on the countryside of his birth to which he had returned to live in later life, calling the entire compilation *The Contrary Experience*.

At first sight *The Innocent Eye* appears to be little more than a physical description of the surroundings in which Read grew up during his first ten years. Consisting of thirteen brief chapters, it is a mere forty pages long. Nevertheless it is clear from much of Read's subsequent writing that the childhood years he spent on the Yorkshire farm were the source of all his later imaginative experiences. They were responsible for his sympathy with the Romantics and the theories of art he deduced from a study of their work. His adult love of music he traces back in the autobiography to his childhood desire to play the violin like Fiddler Dick, an annual visitor to the farm. Childhood is the inspiration for his one novel, *The Green Child: A Romance* (1935). Read acknowledges this debt in the second chapter of *The Innocent Eye*: 'Our memories make our imaginative life, and it is only as we increase our memories . . . that we find with quick recognition those images of truth which the world is pleased to attribute to our creative gift.'[10] This Hobbesian identification of memory with imagination has interesting implications for autobiography. On the one hand it recognizes the way in which our memories have been transmuted within the imagination. On the other hand it makes it harder to differentiate a work of memory (like childhood autobiography) from a work of imagination (like fiction).

It is evident on almost every page of *The Innocent Eye* that Read's memories of his childhood have been subjected to a process of imaginative idealization. A near-perpetual summer reigns. Recalling the farm garden, he writes: 'This green retreat, always in memory a place spangled in leaf-flecked sunlight, with ivy-fruit tapping against the small window-pane, has no grosser associations' (p. 25). Staying out all day without turning up for lunch, he comments: 'I imagine we were severely scolded on our return, but such unpleasantnesses do not endure in the memory' (p. 38). This common tendency to idealize one's past, however, amounts to far more than mere nostalgia for the innocence and security of early childhood. As an adult he finds that his memories of his childhood provide a continuing source of inspiration and a means of healing the wounds inflicted by the adult world. In 'A Dearth of Wild Flowers' he asserts that 'a landscape in which we are born and to which we must always return for the release of the tensions born in exile, has [a] mysterious power of reconciliation, of absolution' (p. 338).

One finds buried in *The Innocent Eye*, in apparent contradiction to this idealization of the past, a strong desire to capture its actuality as it was experienced by the young boy in order better to understand his present day self: 'If only I can recover the sense and uncertainty of those innocent years, years in which we seemed not so much to live as to be lived by forces outside us, by the wind and trees and moving clouds and all the mobile engines of our expanding world — then I am convinced I shall possess a key to much that has happened to me in this other world of conscious living' (p. 17). Read's evocation of the physical world in which he grew up draws heavily on appeals to the five senses. He claims that his first memory of the world is of the sound of horse-hooves as the doctor arrived too late to deliver him. He evokes the first sight he glimpses of his bedroom on summer mornings as 'a hollow cube with light streaming in from one window' (p. 16), the feel and taste of 'sweet rough-skinned apples' and pig-nuts engrained with soil (p. 26), and 'the soft slightly sick smell of cow breath' (p. 27). In search of his childhood sensibility Read seeks to construct his early self by reconstructing the physical world in which he began life. Here is the subjective seeking to discover the nature of its subjectivity by a near-exclusive cultivation of the objective.

Yet at the same time Read repeatedly relates the particularities of his youthful world to archetypal patterns of behaviour. For someone bent on evoking his personal 'first sensations' (p. 17), he spends much of his time recalling life's timeless rhythms and repetitions:

The kitchenmaid was down by five o'clock to light the fire; the labourers crept down in stockinged feet and drew on their heavy boots; they lit candles in their horn lanthorns and went out to the cattle. Breakfast was at seven, dinner at twelve, tea at five. Each morning of the week had its appropriate activity: Monday was washing day, Tuesday ironing, Wednesday and Saturday baking, Thursday 'turning out' upstairs and churning, Friday 'turning out' downstairs (p. 19).

This itemizing has the timeless incantatory quality of the Psalms. Then there is the peculiarly enclosed nature of the Vale. 'This basin was my world, and I had no inkling of any larger world, for no strangers came to us out of it, and we never went into it' (p. 15). This has to be an exaggeration. But he needs to picture a world of abnormal innocence in which, he writes, 'everywhere around me the earth was stirring with growth and the beasts were propagating their kind' (p. 23). He lives in a state of intimacy with animals, falling asleep on one occasion with his head resting on a young calf's flank in the cowshed. This Nativity-like scene reminds one that Christ was seen as a second Adam who shared Adam's pre-Lapsarian idyll in a world of beneficient nature. Like Adam before the Fall, Read as a child is without the adult's experience of pity and terror, 'emotions which develop when we are no longer innocent' (p. 22).

The Innocent Eye is an evocation of the Garden of Eden, the state of nature and innocence to which all of us are born and from which all of us are doomed to be expelled. To the adult Read his childhood 'seems like an age of earthly bliss' (p. 149) spent in the Moors 'where God seems to have left the earth clear of feature to reveal the beauty of its naked form' (p. 46). Intimations of his expulsion from this state of bliss erupt in the form of a memory (or is it, he wonders, a dream bred by the imagination?) of the passage of a large, noisy steamroller the boiler of which rested on enormous bellows, and of a nightmare in which he is menaced by shapes resembling airships. For Read his particular expulsion from a state of innocence came in his tenth year when his father's sudden death led to the sale of the farm and his despatch to boarding school near Halifax. Echoing Blake's title, he calls his expanded autobiography, which covers his years at school, university and the Western Front, *Annals of Innocence and Experience*. Yet his expulsion into the world of experience is implicit in the opening chapter of *The Innocent Eye* where he first starts associating his experience of childhood with the archetypal myth of Eden.

Read's largely unconscious use of this myth is the means by which he attempts to effect a reconciliation between the nostalgic and the questing attitudes towards his childhood. A great admirer of Jung, he was largely instrumental in persuading the publishing firm for which he worked as a director from 1939 onwards to bring out an English edition of Jung's collected works. Jung saw the Garden of Eden as the equivalent to childhood unconsciousness. 'As long as we are still submerged in nature,' Jung wrote, 'we are unconscious, and we live in the security of instinct which knows no problems.' Experience, he goes on, 'is the sacrifice of the merely natural man, of the unconscious, ingenuous being whose tragic career began with the eating of the apple in Paradise. The biblical fall of man presents the dawn of consciousness as a curse.'[11] But Jung does not see it necessarily as a curse. For him, as for Read, the Garden stands for 'an ever-present archetype of wholeness'[12] to which the self can return in later life for renewal by a process of re-integration. This interpretation of the myth correlates with Read's own life, with his autobiographical account of it and even with the theories of art which he developed from it. By associating his nostalgic memories of childhood with an archetypal Eden of innocence and wholeness he simultaneously links his personal sense of the unconscious with the collective unconscious of his readers. Eden is an ideal and therefore no longer attainable state the memory of which can still help the grown man in his search for greater wholeness. Access to this state is through memory even if 'memory is a flower', as Read asserts in the last sentence, 'which only opens fully in the kingdom of Heaven, where the eye is eternally innocent' (p. 55). This highly romantic ascription of divinity to human memory illustrates the way in which Read constructs this exquisitely written brief re-creation of the world of his childhood like a prose hymn. In the case of *The Innocent Eye* autobiography takes the form of an extended prayer.

Laurie Lee: *Cider with Rosie*

Like Read, Laurie Lee sees his childhood as an idyll of the past to which he returns in memory for refreshment and renewal. Although longer, *Cider With Rosie* (1959) has the same air of simplicity verging on naivety which mirrors the child's vision of life. Yet he has subsequently written of its technical complexity. The book occupied him for two years and was written three times. For a start there was the whole problem of compression which brought him face to face with the question of autobiographical veracity. He found that in writing a childhood autobiography 'the only truth is what you remember'. Armed with the conviction that 'there is no pure truth',[13] he found the confidence to distil his sisters' conversations over a period of twelve years to a few dozen phrases, five thousand hours at the village school into fifteen minutes' reading time, and a thousand days into 'a day that never happened'[14] in 'The Kitchen', although every incident in it happened on one of those thousand days.

The shape of the book was dictated largely by his purpose in writing it, 'to praise the life I'd had and so preserve it, and to live again both the good and the bad'.[15] Lee was convinced that the end of his childhood 'also coincided by chance with the end of a rural tradition—a semi-feudal way of life which had endured for nine centuries, until war and the motor car put an end to it'.[16] So he planned the book to start with the dawn of his childhood consciousness and gradually to widen its interest to include first his family, then Slad, the village in the Cotswolds where he spent his childhood, and finally the whole Gloucestershire valley in which Slad is situated. This may sound like the progressive abandonment of subjective for some kind of topographical autobiography. But Lee is well aware of the danger of allowing the autobiographical protagonist to become too much of a ghostly presence: 'The autobiographer's self can be a transmitter of life that is larger than his own—though it is best that he should be shown taking part in that life and involved in its dirt and splendours.'[17] So the book ends on a personal note with an account of his first sexual experience with a girl, the Rosie of the title, and with his new adolescent sense of isolation as his horizons broaden beyond the confines of the valley. By ending on this note he anticipates his departure from home at the age of 19, a departure which he vividly describes in his second volume of autobiography, *As I Walked Out One Midsummer Morning* (1969).

Both these factors, compression and form, add a representative, more-than-personal element to Lee's portrait of his childhood which has helped to make the book a bestseller. *Cider with Rosie* is a celebration of the life of nature into which Lee was plunged as a child of three. It relates his early history to the history of the human race. His childhood begins with the experience of primeval man and ends with that of modern man living in a technological society. Deposited at the opening of the book into June grass taller than his three-year-old self, he is terrified by the experience of being lost amidst the seemingly primeval vegetation, 'thick as a forest' in which 'a tropic heat oozed up from the ground, rank with sharp odours'.[18] For the first time in his life he finds himself cut off from the sight of other humans. The howl of terror that he lets out is like the cry of a newborn baby. This second birth that he experiences at the beginning of the book is into consciousness especially of the natural world of the countryside in which he is to spend his childhood. He and his family 'were washed up in a new land' (p. 11) which he sets out to explore, 'moving through unfathomable oceans like a South Sea savage island-hopping across the Pacific' (p. 15). The images Lee uses are insistent in the parallels they draw between early man's experience of life on earth and his own childish exploration of his home and the surrounding countryside which he scarcely left up to the age of nineteen.

It is as if Lee were attempting to describe the evolution of his individual psyche by relating it to those archetypal images which according to Jung we inherit at birth. Jung argues that they constitute a form of racial memory which at the deepest level of the unconscious merges our unique experience with that of collective mankind. 'Childhood is important', Jung wrote late in life, 'because this is the time when, terrifying or encouraging, those far-seeing dreams and images appear before the soul of the child, shaping his whole destiny, as well as those retrospective intuitions which reach far back beyond the range of childhood experience into the life of our ancestors.'[19] For the youthful Lee, the writer-to-be, those images connecting him to the beginnings of civilization are also destined to become the substance of his art and the basis for his essentially pastoral vision of life. Beginning with the landscape of his kitchen, he traces the way in which his childhood horizons gradually widen through exploration and experience. The outside world first manifests itself 'though magic and fear' to this small boy who shares his 'wordless nights' with his mother in one bed. His heartbeats sound to him like the marching of demons and monsters. Old men live in the walls and floors of the house. The family is terrorized by Jones's goat, 'a beast of ancient dream', 'old as a god' (p. 34). The village 'cast up beasts and spirits as casually as human beings' (p. 36). During his first years in Slad Lee experiences the terrors of primitive man. When they experience a drought 'old folk said the sun had slipped in its course and that we should all of us very soon die' (p. 43), a superstitious fear given widespread religious sanction in ancient civilizations from Egypt to Peru. With the end of the drought 'terror, the old terror, had come again' (p. 43), in the form of torrential rain threatening to invade the family home. Terror is personified in the hysterical reactions of his mother whom he subsequently came to realize was exaggerating the situation out of all proportion to its danger. But by the time he had acquired this insight he had also inherited her primitive fear of heavy rain, a good instance of the power that experience during childhood holds over the thinking and reflecting adult.

Lee's images of primeval life occur most frequently in the earlier part of the book. The valley, he writes, had 'been gouged from the Escarpment by the melting ice-caps some time before we got there' (p. 47). Life in Slad is still in touch with the source of its existence, just as his childhood self is still connected to the origins of his species. When he is sent to the village school at the age of 4 he is confronted by hordes of savages, 'wild boys and girls from miles around' who 'swept down each day' like Vandals to rob him of his lunchtime potato (p. 49). In no time he is pinching someone else's apple, thereby simultaneously becoming a member of society beyond his family and a part of the history of his race. As the book proceeds so the images evoke later more civilized phases of human history. His eldest sister who appears as 'a blond Aphrodite' (p. 72) serves as a representative of Greek civilization, while his youngest brother takes him into Christian times, being 'the one true visionary amongst us, the tiny hermit no one quite understood . . .' (p. 74). Towards the end of the book the time scale advances to post-feudal Britain, still enjoying a basically agricultural mode of existence. His uncles, 'the true heroes of my early life . . . were bards and oracles . . . the horsemen and brawlers of another age', yet their 'lives spoke . . . of campaigns on desert marshes, of Kruger's cannon and Flanders mud . . .' (p. 221). The child's natural hero worship is used to bring his development up to modern times and to place him within that context as the representative of the post-war generation.

Lee celebrates his childhood, despite its fears and terrors, as something shared with earlier civilization. The continuing link is nature which he animates and reveres. Repeatedly he employs language to show how he and his fellow villagers merge into the natural rhythm of life. At home all the children 'trod on each other like birds in a hole' (p. 75); they return to the house at night 'like homing crows' (p. 82). His mother 'possessed an indestructible gaiety which welled up like a thermal spring' (p. 148). She is a source as well as a product of nature, the personification of Mother Nature: 'I absorbed from birth, as I now know, the whole earth through her jaunty spirit' (p. 152). Lee's childhood days were dominated by the change of the seasons. Yet because each season is distilled from a succession of years it acquires in the book a sense of timelessness.

Remembering days spent with the Robinson children, he recalls their 'hide-out unspoiled by authority, where drowned pigeons flew and cripples ran free; where it was summer, in some ways, always' (p. 148). Like Read, he makes overt his idealization as if to acknowledge and condone the adult's knowing falsification of the years of his childhood. For as an adult these moments have become timeless and therefore true to the adult's if not to the child's perception of that time.

Written after Lee had settled in London, the autobiography sets his personal loss within the wider context of the loss that Western civilization had sustained in the course of becoming urbanized and industrialized. In 'An Obstinate Exile', an essay written long after he had settled in London, Lee recognizes that there is no going back to the Slad of his childhood, that he is 'cut off from the country now in everything except heart'.[20] He spends the latter half of the book mourning the disappearance of a way of life that nevertheless he chose voluntarily to abandon. Lee here illustrates what Edward Edinger has written of as the need for the adult to distinguish between a longing for the child's life of the unconscious, a state of nature, and childhood 'inflation' as he calls it, where the ego arrogates to itself the qualities of the larger total self. Edinger amplifies this distinction by citing the age-old controversy between man's yearning for the life of what Rousseau termed the 'noble savage' and his revulsion from primitivism, a revulsion such as Montaigne showed in his essay 'Of the Cannibals'.[21] In Lee's autobiography the child's horizons no longer satisfy the growing young man. After the death of the village squire with its symbolic overtones, 'fragmentation, free thought, and new excitements' came 'to intrigue and perplex him' (p. 269). This recognition of his voluntary participation in what he later came to see as a regrettable if irreversible process is what distinguishes Lee's from, for example, H. E. Bates's childhood autobiography, *The Vanished World* (1969). Bates shares Lee's sense of loss of the pastoral world of his early years but fails to acknowledge his own implication in the movement away from that world. For Bates the games they played, the fruit they picked, the days of harvesting were all better than today's. For him the present means drugs, violence, vandalism. His romanticized version of the past is never countered by any recognition that he is not simply a victim of modern 'progress' but an active part of it.

Lee's account of his gradual detachment from the idyll of his country childhood, an idyll which continues to be a source of inspiration as well as of reproach to the mature writer, closely echoes Jung's account of life's natural cycle:

The child begins its psychological life within very narrow limits, inside the magic circle of the mother and the family. With progressive maturation it widens its horizon and its own sphere of influence; its hopes and intentions are directed to extending the scope of personal power and possessions; desire reaches out to the world in ever-widening range; the will of the individual becomes more and more identical with the natural goals pursued by unconscious motivations. Thus man breathes his own life into things, until finally they begin to live themselves and to multiply; and imperceptibly he is overgrown by them.[22]

This happens equally to individual man and to civilization at large, for, as Jung asserts, 'in our most private and most subjective lives we are not only private witnesses of our age, and its sufferers, but also its makers'.[23] This is what Lee recognizes as he prepares to leave Slad, alone, drawn irresistably to London and then Spain. If one turns to *As I Walked Out One Midsummer Morning* one finds that he came to see his departure from Slad as a further birth into the wider world symbolized by his first awakening in the Spanish countryside: 'I felt it was for this I had come; to wake at dawn on a hillside and look out on a world for which I had no words, to start at the beginning, speechless and without plan, in a place that still had no memories for me.'[24] In one sense Spain offers him everything missing from his cosy valley in the Cotswolds – vast vistas, alien customs, and a feeling of freedom with its 'anarchic indifference' to his presence there. Yet in another sense his experience of Spain is conditioned by his childhood immersion in the countryside and reproduces many of the dramas of his youth but in an adult form. Finally it is Spain that carries him out of the primeval past that Slad represented for him into the midst of the present, the horrors of a twentieth-century civil war. His active involvement in that war is a concrete realization of Jung's assertion that 'the life of the individual . . . alone makes history'.[25]

Richard Church: *Over the Bridge*

For Richard Church there is no conflict between the autobio-grapher's nostalgic evocation of the idyll of childhood and his return to his past in search of greater understanding of his present self. Church is convinced that 'the fundamental purpose of all literary practice, and especially of the art of autobiography' is 'the drama of self-discovery'.[26] This shared celebration of egotism, he argues, enables one 'to assure oneself of one's own integrity . . . as against the constant intrusion of the Public Authority into one's private life'.[27] To this extent the reader can also share in the writer's declaration of independence. Herbert Read held a similar view of autobiography in an age drifting towards self-destruction: 'to establish one's individuality', he asserted in the Preface to *The Contrary Experience*, 'is perhaps the only possible protest' (p. 12). Both writers appear to be echoing Freud's conviction that the pleasure principle, which he maintains is dominant in childhood, is fundamentally opposed to the ethos of Western civilization. Despite centuries of conditioning man remains unconvinced. He remains unconvinced, according to Norman Brown, 'because in infancy he tasted the fruit of the tree of life, and knows that it is good, and never forgets'.[28] The product of this conflict, however, seen in a Freudian light, is liable to be adult neurosis against which the autobiographical act can serve as a preventative or corrective.

Richard Church turned to autobiography in his sixties, writing three volumes in all, *Over the Bridge* (1955) covering the first seventeen years of his life, *The Golden Sovereign* (1957) spanning his next ten years, and *The Voyage Home* (1964) ranging over the following twenty years of his adult life. In the third volume he refers to the first two as 'two static presentations, forming a diptych with childhood hinged to youth', and the final volume as 'a more fugal form to suit the flux of mature life'.[29] *Over the Bridge* opens in 1900 when Church was 7. Chapters 3 to 6 revert to his parents' history and his own years of infancy in Battersea. The book ends with his mother's death in 1910, although there are flashes forward in the final chapter to 1915. Throughout the decade on which he concentrates his attention Church led a remarkably cloistered life with his lower middle-class family. His postman father, teacher mother and elder brother, Jack, are the only major figures of this volume, figures beside whom his teachers, friends and first loves remain insubstantial.

On first reading, *Over the Bridge* appears to be describing the effects of class and social mores on the growth of the child. His was a childhood dominated by the close intimacies and restricted means of his parents' station in life which is held responsible for his own constant ill-health and for his mother's death from bronchial asthma. The opening chapters in particular pay considerable attention to the effects of social conditioning on the young child. Weaned later than is normal, Church feels that 'the close hugger-mugger home atmosphere in the child-life of the great masses . . . makes them over-emotional, unadventurous, matriarch-ridden'.[30] An important theme of the book concerns Church's liberation from these limiting circumstances. But Church is both a poet and a prose writer and his autobiography makes use of the skills of both. When one attends to the poetic elements in the book one sees that its principal theme is the evolution of the creative writer, a theme largely elicited by means of poetic symbolism.

The book opens with one such episode. This appears at first to be no more than a detailed reconstruction of a day in Church's seventh year when he and his brother Jack are given an aquarium which they have to carry home across the Thames through the gang-ridden back streets of Battersea. They escape being set on by a mob of urchins only because of a diversion created by a brawling couple. Once installed in the kitchen window, the aquarium hypnotizes the young Church with its underwater life, its 'distant world' (p. 72) in which fish and vegetation play out a drama of their own. It is not until he is talking about the self-hypnotic process of his creative writing that the significance of his perilous journey home becomes apparent. There he refers to 'my journey over the bridge of time into the past, and the setting up of a transparent theatre into whose fluid element I can stare at the living creatures evoked there, almost forgetting that I myself am one of them' (p. 70). The aquarium then is both an actual object in his past and a symbol for the distance and self-containment of his autobiographical evocation of the past. We learn from the third volume of his autobiography that the journey home with this precious object also represents his 'faith in an idea', 'a faith maintained and bodied in work',[31] the work of creative literature. His autobiography has all the artificiality of a glass tank; but it represents a valued entity (his childhood) retrieved only by his hazardous journey into his past.

In a crucial chapter titled 'A Second Birth' the aquarium becomes associated with the artist within the child. There Church illuminates the earlier symbolic adventure with the aquarium by another incident in his seventh year which has similar portentous ramifications for the future poet. He opens with two pages that describe his fascination with the contents of the aquarium, ending these with a speculation:

It may be that the other-world scenery within the glass walls of Jack's aquarium so played upon my imagination that my sluggish wits were at last awakened. I know that about that time . . . the concentration with which I stared into that small tank . . . began to be applied to the rest of the world around me. I saw things with much more particularity . . . (p. 72).

For Lee, Read and Church childhood memory has proved a major source of artistic inspiration. Church then spends the rest of the chapter recounting how at precisely this moment in his life it was found that he was short-sighted and he was given his first pair of spectacles. At the optician's he finds that for the first time in his life letters harden into separate items of the alphabet. Emerging with his glasses on, he is overwhelmed by a world of clarified perceptions. The urgent demand on his senses, he comments, 'was the beginning of a joyous imposition to which I am still responding today, breathless and enraptured' (p. 75). Church transforms this ordinary occurrence into a vital stage in his growth as an artist, removing him from his childhood cocoon of narcissism and forcing him to attend to the world beyond his childish horizons. He has written of this episode that it 'relates to the dawning imagination of the child', 'to the first phase in the process of vision' which will lead him into 'a discipline habitual with unrelenting intensity',[32] the discipline of literature.

An essential part of that discipline consists in reading the literary classics. In the following chapter Church discovers that, whereas previously no amount of teaching could induce him to read even the clock, let alone books, once he can see clearly with the help of his spectacles he can perform both tasks with ease. Read in this light, the entire volume is a description of the way in which he gradually exchanged the self-regarding autonomous existence of an infant for a special relationship with the outer world that destines him to become a creative writer. Earlier in the book he describes an incident in which, still a toddler, he was with his mother in Battersea Park when she was addressed by two upper-class women cyclists. He is instantly aware of the difference in their enunciation, 'infatuated by the roundness and beauty of the words blossoming on the lips of those two strangers' (p. 32). It is not until the final chapter that Church reveals the full significance of this early childhood memory. There at the age of sixteen he picks up his brother's book of Keats's poems and in a blinding flash discovers his vocation as a poet:

The revelation brought up all the reading I had done since I was 7 years old. I saw a reason in that odd interest in words, and the sound of words through the music of speech. I now understood what had attracted my child-mind to the lips of those two ladies wheeling their bicycles in Battersea Park; an incident lost in the past, because children know no past. But now it rose to the surface of my mind (p. 244).

Church borrows from narrative poetry this technique of retrospectively revealing the symbolic import of incidents which are described largely as pure elements of 'plot' when recounted in their chronologically correct position in the book. Put to autobiographical use, the technique mirrors the way the maturing child slowly begins to look in his past life for a pattern which will justify his commitment to the life of the professional writer.

Church's childhood autobiography in some ways parallels Sartre's. In *Les Mots* (1964), or *Words*, Sartre divides his childhood into two phases, 'Reading' and 'Writing', and organizes the two periods of his upbringing around the two concepts. Church describes the development of his talent as a writer in similar terms. But instead of using Sartre's analytical technique he submerges the thematic content in the emotively charged drama of his life as a child. He employs his early adventures in the world of books as a metaphor for his maturation, a metaphor that enables him to make the emergence of his literary commitment seem almost predetermined. Words first entrance him by their pure sound alone. Once he has learnt to spell words acquire a mystery the power of which he transferred to the books he obsessively read throughout his homebound early schooldays. As with words, reading next becomes a life in itself for him: 'It is an understatement to say that I began to read. I stepped into another life' (p. 83). At first his approach to reading is wholly subjective and he is rapt up by 'the sensation, rather than the material that fed it' (p. 87). Here he is indirectly describing his childish response not just to literature but to all of life. Next style obsesses him. On reading the Book of Job, 'a consciousness of the way things were said dawned within my mind, kindling a light never to be extinguished' (p. 88). Language becomes a 'river', then a 'flood', which carries him on its course using him for its own purposes.

This sense of being possessed by words and used by language conforms with Jung's definition of the artist as one whose 'life is ruled and shaped by the unconscious rather than by the conscious will', and whose 'ego is swept along on an underground current, becoming nothing more than a helpless observer of events'.[33] Nevertheless the artist is still subject to what Jung calls 'personal determinants without which a work of art is unthinkable'.[34] Church, for instance, used books in his childhood as a means of defence against the outside world: 'As soon as I put down my book and took off the armour of words, I felt the winds of life blow cold upon my nakedness' (p. 133). What the autobiography shows is the way in which the world of words slowly became for Church the most compelling experience in his life.

It is significant that he began writing poetry after a year of sacrificing all artistic activities to the task of earning money and nursing his mother. During her terminal illness she would rarely allow anyone except him to touch her. Only when he can see that she has given up on life does he make room for the feminine principle of artistic creativity. The work of art is for Jung like the child that grows out of its mother, or rather out of what Jung calls the unconsious 'realm of the Mothers'.[35] Church was unusually closely attached to his mother right up to her death when he was seventeen. She is in a sense his muse, but too all-consuming while alive to leave him the space in which to celebrate her spirit in poetry. Church also makes use of his brother Jack as a character in the psychodrama surrounding his evolution as a writer. Jack represents the masculine principle in the process of artistic creation – conscious, intelligent, critical. A fine musician himself, Jack shares the austere perfection of the Broadwood–White piano he is bought and applies these severe standards to his muddle-headed younger brother. It is Jack who corrects the young Richard Church's spelling and who leaves his critical comment on the budding poet's first composition—'an admonition to accuracy, to a reverence for tradition and an avoidance of eccentricity' (p. 245). As for his father, he is cast in the role of antagonist to the artistic life. He is described as more boyish and altogether more irresponsible than his son Jack. His distrust of innovation, artistic and otherwise, epitomized by his life-long job as a postman, makes him force Church to leave Camberwell School of Art and to enter the Civil Service. It is the father's opposition to Church's love for the mother arts, rather than for his own mother, which momentarily arouses in the young boy a murderous Oedipal impulse to attack his father with the bread knife. The impulse is fittingly checked by Jack, the only member of the family equipped to discern it. Seen in this light it is also inevitable that both Church and his brother should feel it necessary to set up home for themselves and the piano in order to pursue the artistic life unopposed by their uncomprehending father.

Like Read, Church thinks of his childhood as having been lived in a personal paradise. At the same time he recognizes the inevitability of the process of separation from the parents and the Garden they seemingly inhabit. The self that is to grow into the mature writer is the same self that Church sees as detached from its childhood activities, watching as if from without the effects of its actions on others: 'I believe', Church writes, 'that self-awareness, and the faculty for observing oneself in contact with circumstances, begin in the cradle, and that the infant is already acting a part while at the mother's breast' (p. 63). Without that separating and observing self childhood autobiography would be a difficult undertaking. Is not all that the adult autobiographer remembers that which the separated self in the child observed about itself? Separation is experienced as a form of death which is enacted in *Over the Bridge* by the mother's actual death preceded by her withdrawal from and growing indifference to her son who is nursing her. But separation also allows the separated self to be born into a life of its own. Removed first from his mother, then removing himself from his father, Church eventually parts even with Jack, that substitute father figure whose honest intellectuality Church learns to internalize in his own life and work. This is the theme of the second half of his diptych, *The Golden Sovereign*, which re-enacts the earlier volume's motif of separation and individuation.

Over the Bridge appears to offer a particularly satisfying answer to the dilemma facing all childhood autobiographers—how to reconcile the desire to understand themselves better by an examination of their past with the irresistible impulse to colour their memories of childhood in the falsifying light of nostalgia. In writing about the emergence of the artist in his childhood, Church is simultaneously creating a work of art in its own right. His idealization of the past is responsible for his present self-conscious creation. By evoking the mystery and power that words held for him in his childhood and giving them an emotional reality equal to that associated with the members of his family, he is better able to understand the artist who is at the same time creating a work that is the product of the experiences which it embodies.

6

Parents and Children

Childhood is of immense importance to most autobiographies that concentrate on the subjective experience of the writer. The ways in which it has been treated by twentieth-century autobiographers has been seen to reflect the adults' perspective. The few autobiographers who have attempted to recapture the child's apprehension of life, for instance by using the present tense as Denton Welsh does in *Maiden Voyage*, cannot conceal the presence of their adult self if only because that self is responsible for the literary organization and expression of its experiences as a child. Even if Church is right about the presence of an observing and recording self in the child, the way in which a writer sees his early years says more about the self currently interpreting what the child observed and recorded than about this youthful self. There is the common practice of imperceptibly altering the past over the years so that it answers more closely the needs of the present. As I have already shown, a number of comic autobiographies seek to reverse retrospectively the balance of power that originally obtained between a dominant father and his repressed son. Others, like Read's and Lee's, show the adults' need to idealize their pastoral childhood. Yet others, such as Cyril Conolly's or Edward Thomas's, place most emphasis on the influence school had on their mature character. Both G. K. Chesterton and C. S. Lewis select from their childhood those incidents which have subsequently assumed teleological significance in the light of their much later religious stance. Invariably the writer is reflecting adult preoccupations which govern his selective memories of and attitudes to his childhood.

Some autobiographers continue to see themselves in adult life primarily as products of their interrelationship with a powerful parent. Child psychologists are agreed that parents are normally not just dominant but exclusive presences in the young child's landscape. This chapter concentrates on a few autobiographers who have reflected the vital role all parents play in the formation of their children's personalities. It is ironical that what is known as infantile amnesia, what Edward Thomas calls a 'sweet darkness',[1] prevents most adults from remembering the earliest years of their life. It is during these years, according to earlier orthodox psychoanalytical theory, that many of the most formative incidents and influences occur. The infant starts life with a sense of omnipotence. He is said to possess an oceanic feeling of security which stems from his inability to distinguish himself from the outside world, especially from his mother on whom he is totally dependent. Then slowly she becomes the object of his first perception of a world beyond his instinctual control. Yet even when the child does become aware of her separate existence he is said to identify with her in order to be able to continue to exercise his earlier sense of omnipotence over the world.

These early infantile experiences are rarely reflected in autobiographies, which tend to open with the dawn of memory usually at around the age of four or five. Sergei Aksakoff perhaps comes closest to describing those very early stages, possibly because his relationship to his mother continued to be unusually close well beyond the years of infancy. His earliest memories of her confirm the confusion of identity between the child and its mother that is posited by orthodox theory: 'The constant presence of my mother is part of every recollection I have. Her image makes an inseparable part of my own existence, and therefore it is not prominent in the scattered pictures of my earliest childhood, although invariably a part of them.'[2] Otherwise autobiographies are driven to speculate about a state that they can no longer remember. But they can deduce it from what Richard Church describes as the 'absurd clamour of [one's] craving to see Mother' in the remembered portions of one's childhood. 'Not even the lover, drunken on the wine of sex, puts all his universe into the keeping of one person,' Church observes. 'A child does this, knowing no self-confidence, hardly knowing that it has a self' (p. 167).

This alleged sense of identification with first the mother and subsequently the father is accompanied by an attitude of ambivalence towards them which governs all a child's later relations with his parents. It was Freud's view that a child wants both to love his parents and yet to destroy them when they inhibit his instinctual drives. This emotional ambivalence is said to be initially directed towards the feeding mother. Nevertheless, Freud maintains, early on in the child's life the father too 'becomes a model not only to imitate but also to get rid of'. He goes on: 'Thereafter affectionate and hostile impulses towards him persist side by side, often to the end of one's life.'[3] Ambivalence, then, can be part of the autobiographer's remembered or even current experience. One thinks of Ackerley in later life still unable to come to terms with his enigmatic father who concealed his love for his son as successfully as he concealed the existence of his second family. Ackerley recalls: 'I liked him, I got on well with him, but I was not quite at ease with him . . . I did not find it altogether comfortable to look him in the eye' (p. 75). *My Father and Myself* owes its existence and form to this continuation of childhood ambivalence into later life.

Many of the most successful childhood autobiographies owe part of their success to a conscious or—more often—unconscious embodiment of their ambivalent feelings towards their parents in the texture of their narrative. All three autobiographies discussed in Chapter 4 owe much of their comic appeal to a pervasive ambiguity on the part of the writer towards his father. To portray nothing but the good parent, like H. E. Bates, or the bad parent, like John Osborne, is a diminution of the totality of childhood experience. It is usually the result of a didactic bent which can only find unrestricted expression by heavily censoring the contradictory nature of a child's responses to his parents. By way of contrast, Laurie Lee, who has a strong tendency to idealize his mother in just the same way as he does the landscape of his youth, wins our belief for his glowing portrait of her only because on a number of occasions he demonstrates how habitually 'she tortured our patience and exhausted our nerves' (p. 152). The same is true of *Over the Bridge*. The presence of conscious or unconscious ambivalence towards a parent figure characterizes all the autobiographies considered in this chapter.

Another well-known factor further complicates a child's relations to his parents. This is his introjection of idealized parental figures which he turns into the guardians of his conscience, or his super-ego. Conscience, according to orthodox theory, is constructed from what parents profess or aspire to observe rather than from the actuality of their behaviour. Being largely a part of the unconscious, conscience can become irrationally harsh. The classic scenarios goes as follows. Infantile anxiety at the possible loss of a parent's love becomes internalized in an attempt to control it. In the process anxiety turns to guilt, since conscience cannot be evaded as can external authority. True to this pattern an apparently inexplicable sense of guilt surfaces in most childhood autobiographies. Unconscious guilt, as shown in Chapter 2, is responsible for much of the subtext in the autobiographies of Frank Harris and George Moore. Harris could be said to have spent his life trying to assuage a sense of guilt by living up to the impossible expectations of an internalized and idealized father figure.

Normally, however, this process of internalization leads to a gradual separation from the actual parents who recede in importance as other figures enter the child's life. In her uncompleted autobiography Jean Rhys shows just such a shift in allegiance: 'Gradually I came to wonder about my mother less and less until at last she was almost a stranger and I stopped imagining what she felt or what she thought.'[4] Some children, on the other hand, spend their entire childhood under the domination of one parent or the other, unable to supplant the real-life figure of authority by an internalized substitute until late adolescence and then often incompletely. All the autobiographies considered in this chapter portray childhood as a period characterized by conflict with a powerful parent who has a disproportionate influence, if often negative, on the development of the child. In fact the book which stands as a prototype for this sub-genre gives as much space to the father as to the son, combining biography and autobiography to achieve its sense of balance. Edmund Gosse's *Father and Son* (1907) occupies a similar point of major departure in the history of autobiography as does Lytton Strachey's *Eminent Victorians* (1918) in the history of biography.

Sir Edmund Gosse: *Father and Son*

Edmund Gosse first wrote a book-length biography of his father shortly after the father's death in 1888. His death was in keeping with his life, being the result of bronchitis which he contracted from studying the stars on a succession of cold winter nights in expectation of the personal coming of the Lord. *The Life of Philip Henry Gosse. FRS* appeared in 1890. In it Gosse wrote of his father in that curiously ambiguous tone that was to characterize the later *Father and Son*: 'He was not one who could accept half-truths or see in the twilight. It must be high noon or else utter midnight with a character so positive as his.'[5] If George Moore is to be believed it was he who first suggested to Gosse the idea of *Father and Son* after reading *The Life of P. H. Gosse.* 'I missed the child,' he told Gosse. 'I missed the father's life and your life as you lived it together – a great psychological work waits to be written . . .'[6] Gosse demurred at the time on the grounds that too many of the individuals concerned were still alive. Much later George Moore claims to have suggested telling the father's life in the first-person voice of the son. Gosse, according to Moore, agreed that that would overcome the final difficulty lingering in his mind and proceeded to write the book in 1906. Published anonymously in 1907, *Father and Son* was an instant success, running to five impressions within the first year.

As Peter Abbs has argued,[7] the book is both by the nature of its inspiration (Moore's suggested use of the first person) and of its viewpoint (that of the son) autobiographical. It covers the first sixteen years of the life of Edmund Gosse, from his birth in 1849 to his departure from the family home in Devon for London in 1865. An Epilogue briefly continues the narrative for a further five years to his final break with his father in 1870. Both his parents were extreme Calvinists, members of the Plymouth Brethren. The father was a zoologist of repute who became a Fellow of the Royal Society. However his Calvinist beliefs led him to refute the new theories of evolution making themselves felt in the 1850s. His stand brought ridicule on him in scientific circles and left him in a backwater of his profession. Gosse's mother was a writer of religious verse and popular tracts who died of cancer when the son was seven. On her deathbed his mother's dying words sealed the young boy's dedication to God, a task the father proceeded to devote himself to with the utmost vigour, especially after their move to Marychurch in Devonshire where the father took charge of the local band of the Plymouth Brethren. When his schooling ended Gosse left home for London where his father's friend Charles Kingsley found him an opening in the cataloguing section of the

British Museum. But Gosse was still subjected for a further five years to what he calls the daily 'torment of the postal inquisition' from his father (p. 236). He didn't manage to break this extenuated form of parental control until he was twenty-one, when he claimed to have finally freed himself 'to fashion his inner life for himself' (p. 252).

How successfully he separated himself from his father can best be judged by *Father and Son* itself, written in his late fifties. One's first impression is of a remarkably balanced and impartial account of a childhood warped and constricted by an inhumanly severe upbringing. Natural resentment and anger appear to have been supplanted by love, forgiveness and understanding on the part of the ageing and reflective son. The distance and seeming impartiality which he maintains from his painful memories leave the feeling that here is a highly civilized and humane individual who has had to overcome unusual impediments to reach the calm state of mind in which he recalls his earlier life. It is one of the literary achievements of the book that Gosse contrives to establish this perspective on such emotionally charged material. The strategy was essential if he was to win acceptance from the Edwardian public for a story which up to 1907 could be told only fictionally – as in Samuel Butler's *The Way of All Flesh* (1903). Despite a few outraged reviews, Gosse's book won widespread approval judging from its sales during the first year.

However closer examination of the subtext reveals a more ambiguous and simultaneously a more fascinating picture of the relations prevailing between father and son. The book is at one and the same time more subversive of the father's role than Gosse pretends to be the case, and yet reflects a greater continuing dependence of the son on the father figure than its author seems to realize. Both Gosse and his Edwardian readers appear to be communicating at a second subconscious level where his anti-paternalistic feelings are voiced more strongly and found acceptable by his readers partly because of the way in which the father is seen to live on in the son. By embodying within the book those emotionally ambivalent feelings towards his father which, according to Freud, are inherent in human nature, Gosse instinctually struck a chord that reverberated in all his readers.

These deep conflicting feelings towards his father are buried in a narrative that ostensibly seeks to attribute the primitive conflict between father and son to the impersonal forces of historical change, a moment in the evolutionary time scale on which the father turned his back. Both protagonists are presented as passive agents blown in opposite directions by the winds of historical necessity. The Preface characterizes these winds as 'educational and religious conditions which, having passed away, will never return' (p. 33). Chapter 1 places father and son in two epochs, their mutual affection 'assailed by forces in comparison with which the changes that health or fortune or place introduce are as nothing' (p. 35). The son, it suggests, cannot help being carried forward on the stream of modern progress. The book is offered as no more than a 'document', 'record', 'diagnosis' or 'study' which is 'scrupulously true' (p. 33). It quotes at intervals from objective documents such as his mother's diary and his father's notes and letters to his son. Gosse adopts the tone and approach of a naturalist observer – modelled primarily on his father. Gosse, the conscious author, sincerely believed that these wider forces justified the more personal material on which he drew. At one point he catagorically denies that he is writing an autobiography (p. 217), a curious claim unless one appreciates the extent to which he needed to think that his own experiences were only illustrations of larger historical movements. Shortly after publication he informed one correspondent that the core of the book lay in its 'exposure of the modern sentimentality which thinks it can parade all the prettiness of religion without really resigning its will and its thought to faith'.[8] He clearly is genuine in believing that this is the central theme of the book. Four years before his death he informed another correspondent that his aim in the book had been 'to set down a perfectly faithful and unadorned picture of a succession of moral and religious incidents which can, in all probability, never recur'.[9] It is as if he had convinced himself in the course of writing *Father and Son* that this was his subject – not his father in person so much as 'the puritanism of which he was perhaps the latest surviving type' (p. 239). What 'divides heart from heart', he asserts in the Epilogue, is not individual temperament but 'any religion in a violent form' (p. 248). His father just happened to offer the best instance he knew of such religious fanaticism, this implies.

But one has only to continue reading the passage from which the last quotation is taken to realize that this lofty stance allows Gosse to discharge emotions which far exceed the pretext for their release. His diatribe against extreme Calvinism hides from him (though not from an alert reader) his need to release more personal feelings against his father's oppression than he can consciously admit to. He goes on to assert that evangelical religion sets up a 'vain chimerical ideal' which can lead only to the 'barren pursuit' of 'what is harsh and void and negative', and proceeds to call it 'stern', 'ignorant', 'sterile', 'cruel' and 'horrible' all in the same paragraph (p. 248). These emotive epithets are more likely to have originated in what David Grylls has called a surviving animus against the father,[10] an animus which he has long since repressed and now surfaces indirectly. On only one occasion does he admit to a child's patricidal impulse towards his father. Whipped by the father for a misdemeanour he can no longer remember, he experienced such 'a flame of rage', he writes, that he 'went about the house for some days with a murderous hatred' of him locked in his heart (p. 65). Normally one has to look to the subtext and especially to the imagery which constitutes a vital component of it to appreciate how much anger and resentment Gosse unconsciously harboured towards his father.

Gosse was born on the same day as his father received a green swallow from Jamaica. Betraying no special emotion, the elder Gosse recorded both events consecutively in his diary for that date. In the autobiography Gosse repeatedly uses images of birds and plants to present himself as another product of nature whose freedom and growth has been inhibited by his naturalist father's interference. On at least three occasions Gosse compares himself to 'a small and solitary bird, caught and hung out hopelessly and endlessly in a great glittering cage' (p. 167), 'a bird fluttering in the network of my Father's will' (p. 232), ' "snared" indeed, and limed . . . like a bird by the feet' (p. 242). By extension the image turns the father into a cruel captor who heartlessly inhibits another's life force for his own ends. The same charge emerges from Gosse's comparisons of his situation to a plant, growing on a rocky ledge with just sufficient soil to flower alone on the rarified heights (p. 44), or to one 'on which a pot has been placed, with the effect that the centre is crushed and arrested, while shoots are straggling up to the light on all sides' (p. 211). What is worse, although his father was aware of his contorted growth, 'all he did was try to straighten the shoots, without removing the pot which kept them resolutely down' (p. 211). If the pot represents religious fanaticism then it has been placed there in the first instance by his father and mother.

Read metaphorically, these passages reveal that Gosse's subconscious charge against his father is that he made of his son an extension of his ego. The victim of an overdeveloped moral conscience, the father unconsciously spread his load of guilt among those most subject to his influence, pre-eminently his son. Because he was convinced that his failure to win approval for his anti-evolutionary book was due to some hidden sin within himself, 'act after act became taboo, not because each was sinful in itself, but because it might lead others into sin' (p. 107). Yet behind his father's evangelical zeal lurked a streak of vanity which expressed itself by seeking the advancement of his son. The most obvious example of this is his insistence on having the ten-year-old boy baptized and admitted to communion as if he had attained an adult's state of grace through conversion. A letter written by Gosse at this time refers to his public baptism as 'an initiation into every kind of publicity and glory',[11] which suggests that the son had at least temporarily caught from his father his narcissistic attitude towards the ceremony. In the book after the ceremony the father indulges his son quite uncharacteristically, only warning him against spiritual pride. But it is personal pride that is responsible for the father's usurpation of a child's customary privileges, depriving him of pictures, all forms of fiction and, above all, companions of his own age.

Adding to the father's power over his son is the fact that he is a fanatical minister of an extreme Protestant sect. As such his authority becomes identified in the young boy's mind with the Divine Will whose guidance the father is for ever seeking. Even when Gosse was in his late teens his father would demolish him in argument by assuming that 'he had private knowledge of the Divine Will' (p. 264). No wonder that earlier the six-year-old Gosse 'confused him in some sense with God' (p. 56). In *Totem and Taboo* Freud argues that religion originated in the murder of the father of what Darwin called the 'primal horde' by his sons. Filled with guilt, they proceeded to show 'deferred obedience' to the murdered father by banning murder and incest. Freud concluded that 'the ambivalence implicit in the father-complex persists in totemism and religions generally. Totemic religion not only comprised expressions of remorse and attempts at atonement, it also served as a remembrance of the triumph over the father.'[12]

Faced with a father who both plays God and yet celebrates God's murder by administering communion, Gosse stood little chance of asserting his own instinctual needs. The father had appropriated all the possible roles of father and son.

To pursue Freud's argument one stage further, he sees religion as the adult's response to his sense of helplessness. As a child his helplessness aroused the need for protection through love which the father provided, 'and the recognition that this helplessness lasts throughout life made it necessary to cling to a father, but this time a more powerful one'.[13] This desire to cling to a divine father figure is in Freud's opinion 'patently infantile',[14] an illusion that fulfils a child's wish for protection. By constantly invoking God, Gosse's father is in this view both behaving like an infant himself and inhibiting his son from developing beyond an infantile dependency on him. Faced with a doubly potent father figure, the young Gosse spent much longer than is normal replacing this external embodiment of authority by an internalized one. At five or six years of age he shows none of the guilt feelings that are said to accompany the formation of a super-ego which normally occurs at a much earlier age. For example he is untroubled when he gets away with puncturing his father's lead pipe undiscovered. This accords with Freud's assertion that the very young child lacks any sense of guilt. All it can experience, says Freud, is 'a fear of loss of love, "social" anxiety',[15] which only applies if his misdemeanour is discovered.

At the age of five or six Gosse is still identifying with the father's powers, trying to reproduce them by 'an infantile species of natural magic' (p. 60) that involves bringing dead animals back to life and other such fantasies which 'approached the ideas of savages at a very early stage of development' (p. 61). Even more illuminating is the six-year-old's attempts to flout his father's and God's commands by praying to a wooden idol – a chair. When no divine punishment follows, his confidence in his father's knowledge of the Divine Will, already shaken earlier, is further undermined. But instead of following the usual pattern of introjecting parental authority in the form of a super-ego, Gosse invents an alter ego, as Roger Porter has shown. Porter points out how 'the twin roles of this second self – analytical and aggressive – suggest Gosse's desire both to imitate his father and to transcend him by expressing a forbidden mode of behaviour'.[16] But what Porter does not say is that these patterns of behaviour conform to a psychosocial phase of development normally associated with a one–three-year-old child. At the age of nine Gosse is still seeking to identify with his father by writing childish imitations of his father's monographs on seaside creatures. The father's constant supervision is responsible for his son's retarded development and for his emotional ambivalence towards this patriarchal figure. Right up to the end of his son's adolescence the elder Gosse continues to suspect, like Job, that his son might have sinned, rejoicing in the excuse this gives him to come nearer to his God, supposedly on his son's behalf.

Equally Gosse is subconsciously driven to express the other side of this ambivalence towards his father in his autobiography. Patricidal fantasies are matched by a desire to earn his father's love and approval that survives even the actual father's death. What else impelled him to write not one but two books about his father? It is highly significant that Gosse's memory fails him whenever he breaks out of what he calls their 'Calvinist cloister' and temporarily becomes a normal child surrounded by other children or adults happy to allow him to pursue a child's usual preoccupations. After his mother's death he spends three idyllic months with relatives in Bristol, 'relapsing, to a degree that would have filled my Father with despair, into childish thoughts and childish language'. But, he admits, 'of this little happy breathing-space I have nothing to report' (p. 84). During his eleventh year his new stepmother encourages him to become one of a gang of eight to ten local boys of his own age. Gosse notes how once again his memory of that summer spent in their company loses all distinctions. 'I have no difficulty in recalling, with the minuteness of a photograph, scenes in which my father and I were the sole actors within the four walls of a room, but of the glorious life among wild boys on the margin of the sea I have nothing but vague and broken impressions, delicious and illusive' (p. 157).

The theatrical metaphor in the last quotation highlights the fact that for Gosse his solitary encounters with his father had a glamour and gave him a sense of his own importance as leading co-actor besides which his life outside the home seemed pale and insignificant. Was it not his father's ban on works of imaginative literature that lent them such excitement that he felt compelled to make his living from them as an adult? At the same time he recounts with unbelievable indulgence his father's reason for not agreeing with a fellow member of the Plymouth Brethren that Shakespeare was a lost soul in hell. How are we to know, the father tells his delighted son, that Shakespeare was not converted late in life? Unable to maintain his separate adult identity from that of his father, Gosse within a paragraph assumes full responsibility for this blatant piece of sophistry, claiming that 'it was with a like casuistry that I condoned my other intellectual and personal pleasures' (p. 224). His first attempts to write poetry took the form of pietistic verse, the only form likely to prove acceptable to his father. Nor was his break with his father as decisive as the book makes it seem by positioning it on the final page. As Gide and others have testified,[17] Gosse continued to imitate his father long after his death, assuming the same authoritarian and puritanical attitudes, showing the same discomfort in company, the same tendency towards pedantry in his literary writings. But it is not really necessary to cite such biographical evidence to show how dependent he had grown on his father's actual or remembered company. But for his piety, Gosse concludes, his father would have been 'a charming companion', 'a delightful parent' and 'a courteous and engaging friend' (p. 249). Gosse seizes on this dichotomy to express his own ambivalence towards his father, exposing extreme evangelicism to ridicule so as to be able to preserve the charming companion of his childhood for his life-long admiration and love. It is this half-buried emotional content, complex and paradoxical, that imbues *Father and Son* with its true fascination.

One would expect to find an interesting parallel in Patricia Beer's childhood autobiography, *Mrs Beer's House* (1968). In her case it is the mother who is the dominant parent, bringing her two daughters up in rigid conformity to the dictates of the Plymouth Brethren to which she belonged. Like Gosse, Patricia Beer was faced with a doubly potent parent who didn't hesitate to inform her daughters that Jesus was grieved whenever they showed signs of taking an independent line. Although allowed to read novels, the two girls were never allowed to choose their own books, nor to cut their hair short, and they were barely allowed to play with other children. The family kept itself to itself and provided a hot-house fascination as was the case with Gosse. Patricia Beer acknowledges the power of the spell her mother cast over both daughters: 'I realize it must have been bad for Sheila and me never to run wild, never to be free from supervision and organization, but a great deal of gaiety and drama and company came to us from Mother's management, and on the whole I liked having my development stunted in this way, though obviously I did not consider the situation in these terms.'[18]

The parallels between the books are extraordinary. Even the images show a similarity, as when, for instance, Patricia Beer records that 'we were as much under her control as though we had been on the lead',[19] a simile Gosse also employs of his own ties with his father. Yet *Mrs Beer's House*, good as it is, lacks the compulsive power that *Father and Son* exercises over the reader. To be more precise, it lacks the emotive subtext which makes Gosse's book smoulder beneath the cooling ashes of his retrospective self-control. Despite the way her mother uses both daughters as an extension of her ego, bending them to her will with a skilful appeal to their inherent feelings of guilt, Patricia Beer shows none of Gosse's subconscious resentment. At one point she goes so far as to speculate whether secretly her mother 'was as irked by our slavish obedience as we, secretly, were'.[20] Perhaps the daughters' lack of a father with any semblance of authority left them unconsciously grateful to the mother for fulfilling his function. Besides, the great difference between Patricia Beer's childhood and that of Gosse is the presence of her sister, a natural ally, who inevitably defused the tension that accompanied Gosse's one-to-one relationship. Patricia Beer's story is unusual and well told, but it is a story of 'we' and 'she' with 'he' on the sidelines, not of 'I' and 'he' locked in a conflict that is still being fought out at a sub-conscious level in the autobiography long after the father had died.

W. B. Yeats: *Reveries over Childhood and Youth*

The day after completing his autobiography of childhood in December 1914, Yeats wrote to his father in New York: 'Someone to whom I read the book said to me the other day: "If Gosse had not taken the title you could call it *Father and Son*." '[21] Yet Yeats tantalizingly delayed sending his father a copy of *Reveries over Childhood and Youth* until February 1915, a month before it was published. This ambiguous posture of the 49-year-old son suggests how mixed his feelings towards his father were at the time he wrote the first volume of his autobiography. Most commentators on *Reveries* have tended to simplify the complex web of relationships, past and present, between father and son into a history of the son's revolt against his father. In fact *Reveries* is as much about the son's continuing dependence on the father up to the time of writing as it is about his repeated efforts to find an identity and voice of his own. Richard Ellmann is one of the few critics to point out the way in which Yeats's ambivalent feelings towards his father affected his autobiography. But Ellmann's passing remarks at the beginning of his biography indicate that in his view Yeats's 'autobiographical muse enticed him only to betray him, abandoning him to ultimate perplexity as to the meaning of his experiences'.[22] For me the perplexity is what imbues *Reveries* with much of its fascination. The book, both in structure and narrative content, reflects the confusion of adult narrator and childhood protagonist alike. If the bewildered child is father to the perplexed adult, the adult narrator is also father to the child of his composition, and this interplay adds to the density of his self-portrait.

Yeats wrote *Reveries over Childhood and Youth* between January and December 1914. The book consists of thirty-three short sections covering the first twenty-one years of his life, from 1865 to 1886, although the penultimate section jumps forward to 1892, the year in which both his maternal grandparents died. The narrative is loosely organized on chronological lines, although frequently place supersedes chronology in the ordering of material. Joseph Ronsley has suggested that Synge's unpublished autobiographical notes in Yeats's possession and the appearance of Joyce's fictional autobiography, *Portrait of the Artist as a Young Man*, in serial form starting in February 1914 might well have 'awakened Yeats to the possibilities of disconnected narrative'.[23] Yeats himself described it as 'less an objective history than a reverie'.[24] Its dream-like passage from one memory to another, its highly personalized sense of Irish history in the making and its mesmeric prose style make it a departure from the traditional approach to childhood autobiography and establish it as a unique contribution to the genre.

After *Reveries* was published in 1916, first by his sister's Cuala Press, then by Macmillan, Yeats went on to write further volumes of autobiography. Of these only the 1926 edition of

Autobiographies need concern us here. This edition included *Reveries* and *The Trembling of the Veil*, an account of his life between 1886 and 1896. In the process of bringing together these two consecutive accounts of his first thirty-one years Yeats eliminated overlaps and redundancies (including passages on Dowden, O'Leary and Taylor) by deleting a few passages from *Reveries*. None of these eliminations is of major significance, so I will confine my attention to the text of *Reveries* which is most accessible today, that contained in Macmillan's English edition of *Autobiographies* (1955). Although it has been argued that both *Reveries* and *The Trembling of the Veil* constitute a unity, Yeats himself provided the justification for considering *Reveries* as a separate entity. After informing his father that he had finished *Reveries*, having brought them down to 1886, he continued: 'After that . . . they would have . . . to be written in a different way. While I was immature I was a different person and I can stand apart and judge.'[25]

In 1914, the year he wrote *Reveries*, Yeats was also busily engaged in collecting his father's letters together in order to have a selection of them published by his sister's Cuala Press, a project realized in 1917. The extent to which the son's identity was still confused with that of his aged father can be gauged from the letter he wrote his sister after completing *Reveries*: 'I have written it in some sort as an "apologia" for the Yeats family and to lead up to a selection of our father's letters . . .'[26] Again, just before publication date Yeats, in excusing himself to his father for not commenting on the lengthy letters his father had been sending him, wrote: 'To some extent you must take the book of Memoirs as a comment. When I was writing it I thought constantly that you would write me, after it . . . letters about it.'[27] Even the fact that these explanations constitute a verbal smoke screen released for defensive purposes only further emphasizes the extent to which the presence of his father still haunted Yeats while writing *Reveries*. Far from writing his father out of his system in his childhood autobiography, Yeats reveals the continuation into later middle age of his inability successfully to separate himself from his father's dominating personality.

Although Yeats shows signs of emotional ambivalence in the sense that Freud used that term, his prolongation of this ambivalence into later life also suggests the potent presence of what Jung called the father-imago or archetype. According to Jung the child's relation to his father offers an 'infantile channel along which the libido flows back when it encounters any obstacles in later years, thus reactivating the long-forgotten psychic contents of childhood'.[28] But instead of recognizing that they are long-repressed memories from childhood the adult interprets these infantile relics as 'feelings of being secretly guided by otherworldly influences'.[29] These influences can be attributed to a divine or daemonic figure (the archetype of the father). So that 'the personal father inevitably embodies the archetype, which is what endows his figure with its fascinating power'.[30] Because the archetype of the father is essentially ambiguous (good or evil daemon) it is capable of having diametrically opposite effects, eliciting from the son fear and love simultaneously, so that he is left saying yes and no at the same time. This is a fairly close diagnosis of the position Yeats finds himself in the course of writing *Reveries*.

The first twenty pages of the book are confined to his memories of Sligo in the north west of Ireland where he stayed with his maternal grandparents, the Pollexfens. To read *Reveries* is to gain the impression that Yeats spent most of the first nine years of his life with his fiery grandfather whom, he says revealingly, he confused with God. He actually describes Sligo as a Garden of Eden from which he was expelled. The cause of his expulsion was his father with whom he was forced to spend the rest of his childhood and youth wandering over the face of the earth, or at least vacillating between London and Dublin. The biographical facts are very different. Yeats spent the first two years of his life with his parents in Dublin and the next five with them in London. During this period he spent holidays most years with his grandparents but lived with them apart from his father for only less than two and a half years between the ages of seven and nine. Yeats's exclusion of his father from his account of his childhood up to the age of nine is a form of wish–fulfilment, a retrospective attempt to minimize the extent of his father's influence over him during his formative early years. In fact his father was extremely anxious about leaving his eldest son in the care of his parents-in-law in 1872 and showed his distrust by leaving firm instructions that he was to be 'made more robust by riding or other means – *not by going to school*'.[31] Yeats does record in *Reveries* the one visit his father made to Sligo during this period when he took it upon himself to teach his son to read, employing terror to conquer his son's wandering mind. Nevertheless with the help of these tactics he built the foundations for Yeats's adult identity as a practising poet. Not only did he build the foundations but he proceeded to construct much of the rest of the building, introducing his son to poetry, encouraging him to try his hand at writing prose and verse, and imparting to him a philosophy based on a belief in artistic freedom of which Yeats could not properly rid himself without abandoning his chosen profession.

Behind his father, I would suggest, lay the archetype of fatherhood embodied in the artistic daemon. The fact that the father was himself an artist who had rebelled against the conventional life of his day made it doubly hard for the son to find an outlet for his own rebellion. Yeats struggled hopelessly against this potent combination. Artistic freedom for the father meant freedom from dogmas of all kind, especially religious dogma. The son spent much of his life searching for a satisfactory substitute for a belief in Christianity which his father's agnostic indoctrination put beyond recovery. The incident Yeats cites during the 1872 visit of his father shows him first desolated by his father's unbelief, then comforted by a child's belief that the newborn were a gift of God, and finally plunged once more into the gloom of his father's scepticism on learning about the facts of sexual reproduction. This first failed attempt at rebellion against his father's influence typifies all Yeats's later and more desperate efforts at throwing off his father's domination. Next he is informed by an aunt that his father is about to take him away from Sligo to live in London where he would be 'nobody at all'.[32] Yeats indirectly charges his father here and elsewhere in the book with infecting him with his deracination and so making him feel like a stranger at school in London. Yet subsequently Yeats undermines this charge by ascribing positive value to the European influences to which his father introduces him and which save him from succumbing to the narrow-minded Irish nationalist views on literature and politics to which he was subjected in his late teens.

All Yeats's revolts turn out to be half-revolts, token gestures of defiance, although much of the time he does not appear to be fully aware of the way he undercuts his own acts of independence. Taken to see *Hamlet* by his father, he records how he identified with Irving rather than with Ellen Terry whom his father idolized. But he goes on to say how for many years after he modelled himself on the heroic self-possession of Hamlet, the character to whom his father first approvingly introduced him. Similarly his solitary nights spent out of doors in his teens are modelled on *Walden* a passage from which his father had read to him shortly before. He refuses to go to Trinity where his father and his forebears had gone before him, but not, it turns out, in any spirit of defiance, simply because he knew that his classics and mathematics were not good enough – a fact that he dared not disclose to his father at the time. Instead he attended art school in Dublin where, he discloses, 'my father, who came to the school now and again, was my teacher' (p. 79). Then there is his account of his two-year friendship with his father's one-time friend, Dowden. To begin with Yeats glories in his admiration for Dowden when his father had already become alienated from his ideas. Once again his independent stand is quickly undermined by a reference on the next page to some of Dowden's 'conventionalities and extravagances that were, my father and I had come to see, the violence or clumsiness of a conscientious man hiding from himself a lack of sympathy' (p. 87). Father and son are once more united against the outside world.

Yet Yeats still seems unaware of this volte-face a year after he had finished *Reveries* when he wrote to his father that the one section of the book that he feared might offend him was that dealing somewhat harshly with Dowden. Dowden, he explains, cannot be left out, 'for, in a subconscious way, the book is a history of the revolt, which perhaps unconsciously you taught me, against certain Victorian ideals'.[33] When one bears in mind the way in which his father continued to love Dowden, the man, after he had grown disenchanted with Dowden's outlook, one realizes that Yeats has used Dowden as a surrogate for his father. Dowden was a figure whom it was safe to condemn and who was dead, unlike his father who was all too alive and anxiously waiting to read his son's version of their early life together. In fact the more one examines *Reveries* and documentation relevant to it the clearer it becomes that Yeats, the narrator of 1914, still harbours the same ambivalent feelings about his father that the child and youth in the book felt. There is clear evidence, for example, of self-censorship in matters relating to his father if one compares *Autobiographies* to the

Memoirs he wrote with no intention of having them published. In the *Memoirs* he relates an incident in which his father quarrelled with him so violently that he smashed the glass in a picture frame across the back of his son's head,[34] something that Yeats omits to mention in the published version. He exercises the same restraint when summarizing the plot of his early verse play, *Love and Death* in Section XVIII of *Reveries*. His précis of a play based on one of his father's drawings significantly omits to mention the central event, patricide. This is a theme he reverts to again and again in his imaginative writings throughout his life. His last attempt at killing off this father figure was in *Purgatory*, a play written shortly before his death.

The closest Yeats comes to defining successfully the difference between his own youthful ideal of literature and his father's model comes near the end of *Reveries* when he opts for 'personal utterance' which his father dismisses as 'only egotism' (p. 102). Even then he is quick to qualify this distinction by admitting that the early poems he wrote in pursuit of this ideal strike him now as 'little but romantic convention, unconscious drama' (p. 103). Dramatic statement was the mode his father championed. Yeats concludes wistfully: 'It is so many years before one can believe enough in what one feels even to know what the feeling is' (p. 103). *Reveries* makes it clear that the difficulty stems from his artist father whose feelings become confused with the son's in the latter's unconscious. As Jung sees it, the danger for the son is a powerful unconscious identity with the archetype of fatherhood: 'Not only does it exert a dominating influence on the child by suggestion, it also causes the same unconsciousness in the child, so that it succumbs to the influence from outside and at the same time cannot oppose it from within.'[35] This appears to be Yeats's position both in childhood and at the time of writing *Reveries*. No wonder the autobiography ends on an enigmatic note of frustration bordering on despair: 'All life weighed in the scales of my own life seems to me a preparation for something that never happens' (p. 106). Yet ultimately it is Yeats's frankness in portraying the history of his ambivalent relations to his mesmeric father that makes of his confusion a compelling and moving autobiography. In both Gosse's and Yeats's autobiographies the presence of unresolved psychic conflict in the subtext produces a tension between intentional and unintentional levels of meaning that give them the semblance of life's complexity.

Osbert Sitwell: *Left Hand, Right Hand!*

The childhood autobiographies of both Gosse and Yeats can be seen as attempts at writing their dominant fathers out of their systems. Viewed in this light both books could have been considered in the next chapter which focuses on autobiography as self-therapy. Osbert Sitwell's five-volume autobiography, *Left Hand, Right Hand!*, is poised between the perspectives of this chapter and the next, a book that is simultaneously obsessed with the influence of the father on his son and concerned to lay the ghost of this larger-than-life baronet. The design of Sitwell's autobiography is ambitious and complicated. The first volume, from which the title of all five is taken, was written before the death of Sir George Sitwell in 1943 but not published in serial form in the first instance until 1944. Therefore it may bear revisions made after the father's death. *Left Hand, Right Hand!* (1945), the first volume, concentrates on the first nine years of Osbert Sitwell's life from 1892 to 1901. *The Scarlet Tree* (1946) spans his schooldays from the age of ten to eighteen. *Great Morning* (1948) is confined to his pre-war army career during 1912 and 1913 and to his introduction to high life and art circles in London. *Laughter in the Next Room* (1949) takes the story from 1914 to his father's death in 1943. *Noble Essences* (1950) forms a coda in which Sitwell offers a gallery of portraits which help delineate both himself and the era covered by the other four volumes. All five volumes make free with strict chronology in furtherance of wider artistic considerations.

The scope and intentions of this massive autobiography go far beyond the concerns of the present chapter. In his introduction to the first volume Sitwell makes clear that the sequence is a life-and-times, as much a history and multiple biography as it is an autobiography. He explicitly warns the reader how much of himself he intends to exclude: 'I will not . . . concern the reader with those parts from which I want, myself, to escape, for the aim of this book is to beguile, and not to improve, the mind.'[36] This evasive strategy has led commentators to dismiss the autobiography as mere reminiscences which, in Stephen Spender's words, 'are revealing of his family, but tell us little about what it feels like to be in Sir Osbert's skin'.[37] Although it is true that the portrait of his eccentric father tends to overshadow his self-portrait, I would argue that the son's delineation of his relations to his father, not just in the past but in the way he narrates in the present events of the past, reveal more about the autobiographical subject than Spender allows or than Sitwell himself intended. I will be confining myself exclusively to this crux for present purposes.

Osbert Sitwell suffered almost as much as did his sister Edith at the hands of their father, even if he had the compensation she lacked of his mother's life-long love. For a start Sir George, according to his eldest son, considered sons, especially sons when they were small, as 'a valuable extension of his personality' (p. 94), and only 'loved children until they were old enough to reason, express their views and show a will of their own' (p. 192). Although the father is offered to the reader as an amusing eccentric it becomes clear from Osbert's description of him that he was a neurotic introvert, the type who, as Jung describes it, 'shuts himself up with his complexes until he ends in complete isolation'.[38] In fact the father conforms in almost every respect to Jung's description of the extreme introvert: turning his back on the present to inhabit in his case an imaginative reconstruction of the Middle Ages; mistrustful of everyone around him including his wife and children; frugal in all matters except his own extravagant expenditure on his various properties; surrounding himself, as Jung puts it, 'with a barbed-wire entanglement so dense and impenetrable that finally he himself would rather do anything than sit behind it'.[39] True to type the father suffers a nervous breakdown when he realizes that he is unable to get his way in everything and is forced to face the fact that in some matters he is in the wrong.

Looked at objectively, then, Sir George's state of increasing neurosis led him to behave towards his son with at times unbelievable inhumanity. Because Osbert was the eldest son and therefore dependent on his father for his income and expectations his father was able to dictate to him his life-style and profession with arbitrary whimsicality. The most extraordinary example of this occurs when the father arranges for his son to be gazetted into the Yeomanry, a fact with which the son becomes acquainted by reading about it in *The Times*. Osbert's amusing account of this incident in the autobiography shows no surface resentment at his peremptory treatment. To refuse to join his regiment at Aldershot would have rendered him guilty of mutiny, he argues. But he then spends many pages depicting the misery of his life at the barracks which he represents as an experience worse than boarding school in its abhorrence of anything intellectual or artistic. His state of acute misery and his amnesia about much of his time at Aldershot are the direct consequence of his father's arbitrary action. Yet his only mention of him during this episode in the autobiography concerns petty quibbles over his allowance. Eventually Osbert staged the first of his semi-breakdowns, ran away to Italy where his father was living at the time and endured a series of confrontations over his future career with his impossible parent. His father would repeatedly summon him to his room where he would harangue him from behind the 'cloudy cover' of his mosquito net where he sat 'like the Deity' who 'remained only partially revealed'.[40] The same confusion between the father and God as was experienced by Gosse is present despite the protective presence of irony in the mature Sitwell. So that it is not surprising to find him berating himself at this time in the manner of innumerable Old Testament sinners for incurring his father's wrath. All he had to do, he realized a decade later, was to join in the game his father was playing, although he fails to recognize that it was his young life and happiness that was the subject of this game.

In the end the father agreed to arrange his transfer to the Grenadiers only because he understood the discipline there to be stricter than in the regiment his son so loathed. This ugly incident is in this way rendered seemingly harmless by the all-forgiving son's use of an amused, mildly humorous tone and stance. But behind that benevolent stance lies a bond with the father that combines love and hate and that still dominates his sense of self when he was writing *Great Morning* in his mid-fifties.

Simultaneously Osbert seeks to explain away his own and the rest of the family's increasing misery at this time by attributing it not to its direct source in Sir George, but to the darkening climate in Europe leading to the outbreak of the First World War: 'Some agency was at work,' he writes in *Great Morning*, 'both in the family and outside it, loosening the fibres, darkening the colours.'[41] What he is anticipating is the disgrace that fell over the family when in March 1915 his mother was imprisoned for three months for debts which his father pig-headedly refused to repay. Osbert shows extreme reticence over recounting this episode and spends most of the small space he gives to it explaining why his father acted as he did – to expose and render harmless the fraudulent money-lender who had milked Osbert's mother of huge sums in interest. It never occurred to his father, he writes in *Great Morning*, that his wife might land up in gaol as a consequence of his intransigence. When he reaches the catastrophe in *Laughter in the Next Room* he confines himself to one sentence in which he blames the outcome on his mother's inherited absence of any sense of money or even of simple arithmetic and also on his father's pride as well as his habit of preferring not to see what is likely to prove painful to him. But most of his energy is devoted to describing the 'highly coloured, evil levity of spirit'[42] in the London house and the storms which battered the ancestral home at Renishaw during this climactic time. Once again he makes the perfectly avoidable family calamity appear as a symptom of a much wider malaise. One cannot escape the feeling that frequently he is reacting as if he, not his father, had been responsible for landing his mother in gaol. On the one hand he exculpates his father when it would seem more natural for him to rebel against his high-handedness. On the other hand he depicts his father as a modern Don Quixote, too ridiculous to be taken wholly seriously.

Osbert first adopted this air of detached amusement towards the outrageous behaviour of his father in his late teens. His autobiography owes to this use of comic caricature its brilliant portrait of the half-mad, tyrannical baronet. Osbert Sitwell shares with Gosse and Yeats a need to distance himself from the overpowering presence of his father during his childhood. Like them also, he is unaware of the extent to which the subtext reveals his continuing identification with the figure he thinks he has turned into a harmless comic grotesque. There are, for instance, the occasional revelations of the extent to which he modelled himself on his father and 'believed that his every opinion on every subject must be the correct opinion'.[43] Apart from these, one notices the number of parallels accumulating between the lives of father and son. In *Façades*, a joint biography of the three Sitwell children, John Pearson describes how Osbert, without being fully conscious of it, echoed Sir George in being sensitive, solitary and bookish. His reaction against religion and compulsory games at Eton shows him cultivating similar aversions to those of his father. When the outer world becomes too oppressive for him, as it does at Aldershot barracks, he imitates his father's retreat from life by staging his own virtual mental breakdown.[44]

Where Osbert, like his sister, is persuaded that he most differs from his father is in his pursuit of the arts which he felt provided him with a sense of separate identity. Yet in the first place Sir George, as his son remarks at one point, showed at least a theoretical respect for artists even if he was barely acquainted with any (p. 190). Further it becomes clear that Sir George is a writer manqué, managing to publish only obscure works of mainly antiquarian interest, but with as much ambition to receive recognition as his successful son. According to Jung, 'Nothing exerts a stronger psychic effect upon the human environment and especially upon children, than the life which the parents have not lived.'[45] Osbert's rise to literary eminence, then, proves to be yet another, less conscious form of identification with a father who exercised his extraordinary influence over him up to the time of his death. It is ironic that even in death the son modelled himself on his father, leaving a will that caused the same consternation among the survivors as his father's had done.

What emerges from this brief and highly selective glance at Sitwell's lengthy autobiography is, I suggest, the fact that in portraying his father at such length he is simultaneously portraying much of himself, even more than his own stated belief in heredity would suggest. His extended portrait of Sir George is as central to his autobiographical purpose as is his portrayal of the changing social climate of these years. Both have gone into the make-up of the artist who is determined to make of his autobiography something 'old-fashioned and extravagant', in fact 'gothic' (p. viii). Isn't this the same Gothic world to which Sir George withdrew from a modern age which father and son alike viewed with abhorrence? Behind his therapeutic need comically to distance his father persists the deeper need to acknowledge the extent to which his sense of self is dependent upon a father whose unfulfilled life he spent much of his time living out by proxy.

7

Autobiography as Self-Analysis

There is a widespread belief among creative writers in general and autobiographers in particular that, as D. H. Lawrence claimed, 'one sheds one's sicknesses in books – repeats and presents again one's emotions, to be master of them'.[1] This is very similar to the assumption underlying the psychoanalytic process. The adult, it is assumed, can rid himself of behavioural patterns learnt in childhood and continued into adult life in a disguised or neurotic form only by recalling and reliving those experiences of childhood which produced the defensive responses in the first place. Numerous autobiographers are motivated by a similar desire to free themselves from their past by re-experiencing it. John Lehmann typifies this aspiration at the start of his three-volume autobiography: 'I had come to the point where I wanted to understand myself by analysing my past.'[2] Yet when one asks oneself how successful Lawrence or Lehmann have been, one remembers Lawrence's claim in mid-life that he would have written a different *Sons and Lovers* then to that in which he earlier believed he had mastered his childhood emotions, while Lehmann defeats his quest for self-knowledge by the omission of large areas of his private life. Besides it has already been shown that the past is irretrievable in its pristine form once it becomes subject to adult reflection and that it is further distorted during its embodiment in a written narrative which assumes a life of its own. Memory itself distorts. The present-day observing and recording self frequently imposes its immediate needs on the material of the past. Besides what is that observing and recording self which the autobiographer draws on to improve his understanding of himself? Is it not also the self he hopes to improve? If so, how can he hope to observe it? By 'turning his text back upon itself', perhaps, as Elizabeth Bruss suggests, 'to examine the vantage point rather than the view'?[3] But who is it then who is turning it back? Besides, as Bruss points out, the likely outcome of this much-used procedure is a realization by the autobiographer of his failure.

Clearly there is a fundamental conflict implicit in an autobiographer's desire to understand and so to be more in command of the self and in his unconscious desire to protect his ego by turning his life into a shape that is interesting and acceptable to himself and the world. This is especially true of the confessional mode in autobiography which has always shown a vested interest in justifying the current reformed self while sacrificing a past pattern of behaviour which no longer conforms to its adult needs. 'Sick of being a prisoner of my childhood,' Clive James writes at the start of his autobiography, 'I want to put it behind me.' To do this he has to satisfy his confessional urge, and, he argues, 'the mainstring of a confessional urge is guilt', guilt over all the crimes he has 'mainly been successful in not recollecting'.[4] But a further difficulty arises here. In dragging those crimes into the light the writer is looking for some form of remittance from guilt. So the confession is likely to be constructed so as to win absolution from the reader.

Like the analyst the reader may be able to discern the falsifications this confessional act tempts the confessor to commit, but there can be no corrective interaction as in a therapeutic situation. Besides, is it really the reader to whom the writer confesses, or himself in public? If himself, how is he to avoid circularity, self-deception and the reinforcement of unconscious repressions which neither the writer nor the writer as reader has an interest in exposing? The odds, therefore, are stacked heavily against the writer who turns to autobiography as a form of self-therapy.

Freud himself underwent a period of self-analysis which helped him break through to a partial understanding of the workings of the unconscious. On the other hand Freud subsequently concluded that 'in self-analysis the danger of incompleteness is particularly great. One is too soon satisfied with a part-explanation, behind which resistance may easily be keeping back something that is more important perhaps.'[5] Jung concurred: 'As regards ourself we remain blind despite everything and everybody.'[6] In more recent times Karen Horney and others have maintained that under the right circumstances self-analysis can be fruitful. But the procedures for defeating the ego's defences have to be rigorously pursued. They cannot coexist with the requirements of a literary narrative. Many autobiographers for whom self-analysis is the most important motive in their undertaking acknowledge this incompatibility by incorporating it into the fabric of their narrative and use paradox in particular to give expression to this intractible conjunction. 'I write down my story and state my present problem', H. G. Wells announces at the beginning of his autobiography, 'to clear and relieve my mind. The story has no plot and the problem will never be solved.'[7] So that as well as the compulsion to write about the troublesome past there coexists in autobiographies of this kind a gaunt courage in setting out in search of a goal which the writer knows will elude him. Yet he can at least construct a monument to commemorate his defeat. Maybe it is the acceptance of defeat that constitutes his reward.

This therapeutic urge to undertake the impossible, to write out one's sickness, is probably most prominently in evidence in those autobiographies of the 1920s that were written by men who had experienced the horrors of the Western Front during the First World War. Both Robert Graves and Siegfried Sassoon suffered a mental breakdown during the course of the war. They and Edmund Blunden found themselves reliving their lives in the trenches long after the end of the war, in nightmares and hallucinations, in their poetry, and in their autobiographies. Right at the outset of his autobiography, *Undertones of War*, Blunden, true to form, acknowledges the delusive nature of his compulsive journey back into the past.

I must go over the ground again.
    A voice, perhaps not my own, answers within me. You will be going over the ground again, it says, until that hour when agony's clawed face softens into the smilingness of a young spring day . . .[8]

His answer to the predicament he finds himself in is to posit two voices, each holding opposite views of the undertaking. The two voices sound very like a dialogue between conscious and unconscious selves. All three writers seek in their autobiographies to lay the ghost of mass destruction in which they played a part. In the course of writing about their past all discover within themselves the destructive instinct which had made them want to efface the enemy, efface the 'old men' back home responsible for the slaughter, and, most tellingly, to efface themselves. They were torn between revulsion at the killing they witnessed and an urge to abandon themselves to the futile mass sacrifice. It was a dilemma from which the only escape lay in death, serious injury or mental and physical breakdown. The trauma of the experience was so severe and long drawn out that it took Graves and Sassoon a decade before they felt able to write their autobiographical accounts of it. The different nature of these accounts of a similar experience is due not just to the difference in their personalities and background, but to the autobiographical strategies each adopts in an attempt to reconcile irreconcilable aims and needs.

Robert Graves: *Goodbye to All That*

At the end of the Great War Graves was suffering from nervous prostration. Unable to use the telephone, sick every time he travelled by train, haunted by scenes from his first four months at the front, he retained these daydreams of battle 'like an alternate life',[9] he records, until well into 1928, the year before he wrote *Goodbye to All That*. He had written poems about the war while it was in progress but had destroyed them as too journalistic. During the war he also began a novel about his experience in the trenches which he tried to finish several times after the war. But his motive – 'to rid myself of the poison of war memories' (p. 262) – conflicted with the need to distort those memories to fit a fictional plot. Finally in 1929 he felt strong enough to recall it all in the form of 'undisguised history' (p. 262). The strength of his compulsion to lay the ghost of his military past is evident in the way in which the original version of *Goodbye to All That* was part dictated, part written at tremendous speed between May and July and revised during August of 1929. The book consists of thirty-two short sections spanning Graves's life from his birth in Wimbledon in 1895 to 1926. A final section in the 1929 edition offers a highly selective version of his life between 1926 and 1929. He added a brief Prologue and Epilogue to the revised edition of 1957 which added a few biographical facts concerning his life after 1929.

There is no mention of the irruption of Laura Riding into his life in January 1926, but in the original edition of 1929 Graves did include a Dedicatory Epilogue to her in which he explains that he omitted her from the autobiography because she belongs outside the life it represents. By 'living invisibly, against kind, as dead, beyond event',[10] he writes, she has helped him detach himself from his past life. In October 1929 Graves and Laura Riding left England to settle in Mallorca. *Goodbye to All That* was written in a mood of revulsion at a turning point in Graves's life. His personal relations with his wife and Laura Riding had undergone a crisis ending in divorce and Laura's attempted suicide. He identified European values with the horrors of the war. He wanted to start life anew. As he wrote on the day he finished the autobiography, 'Another month of final review and I shall have parted with myself for good.'[11] The following year he published an 'appendix' to *Goodbye to All That*, *But It Still Goes On*, which included 'material from the same autobiographical files'.[12] Included in this collection is 'Postscript to *Goodbye to All That*', a reprint of an article he wrote for the *Daily Mail* as a general reply to all his critics. The article is a cynical exposé of how in *Goodbye* he 'deliberately mixed in all the ingredients that . . . are mixed into other popular books'[13] – meals, murders, ghosts, famous men, love affairs, and above all battles. His motive, he proclaimed, was to make a lot of money fast in order to provide for his family in England and to finance his new life in Mallorca. This he succeeded in doing. When the book was published in November 1929 it sold 30,000 copies in the first month alone.

In *The Great War and Modern Memory* Paul Fussell quotes at length from this 'Postscript' to show that *Goodbye* was written in such a mood of cynicism that it should not be trusted factually. Rather it is 'a triumph of personal show business'.[14] This is to take Graves too much at his own word. The 'Postscript' was written when he was suffering from an acute feeling of revulsion from English society. What Fussell and others have failed to appreciate is the radical nature of the revision that Graves undertook in 1957. Graves explicitly spells out the kind of changes he made in the Prologue which he wrote for the new edition. But the extent of them is best indicated by the answer he gave to an interviewer in 1968: 'I entirely rewrote *Goodbye to All That* – every single sentence – but no one noticed . . . It's an entirely new product.'[15] To apply the 'Postscript' of 1930 to the revised edition of 1957 which we all use now is anachronistic. In addition Graves offers his own paradoxical definition of the kind of truth that is demanded in any account of war: 'The memoirs of a man who went through some of the worst experiences of trench warfare are not truthful if they do not contain a high proportion of falsities. High explosion barrages will make a temporary liar or visionary of anyone . . .'[16] Here Graves is arguing, not for the novelist's need to lie like truth, but for the pre-eminence of subjective experience in any autobiography centred on the writer's experience of trench warfare between 1914 and 1918.

For Graves the problem was how to tell the previously untellable, how to use conventional narrative techniques to describe the unprecedented horrors of war as he experienced it. In his 'Postscript' Graves writes that the solution he hit upon was to organize his material in the form of a number of short stories or 'situations'. Fussell acutely analyses the way in which Graves cultivates the form of the short theatrical anecdote. These often take the shape of virtual playlets or what Graves refers to as 'caricature scenes'. For Fussell *Goodbye* is 'a satire, built out of anecdotes heavily influenced by the techniques of stage comedy'.[17] In many of the most caricatured episodes Graves represents himself as an anomaly attempting unsuccessfully to conform to absurd social conventions. One remembers the scene behind the lines in France in which Graves in full uniform is squashed into a school desk like an overgrown schoolboy to be lectured by a colonel who has been outraged by overhearing a soldier address a lance corporal by his forename. Or there is Graves, convalescing from his wounds on the Isle of Wight, playing at being a pseudo-yachtsman in the Royal Yacht Squadron smoking room.

Where Fussell tends to view *Goodbye* as a very loosely connected series of dramatic interludes, a focus on the therapeutic content of the book offers a more integrated thematic view of it. The 1929 edition ends with a four-page summary of Graves's life between 1926 and 1929 which, he writes, 'must be read . . . as (a story) of gradual disintegration'.[18] In fact *Goodbye* as a whole is about the process of disintegration, or of a failure in integration. Graves sees his life from its inception as a progression of self-divisive experiences. Even his lineage counted against him. His father was of Irish descent, his mother part German. When it was discovered at Charterhouse that his name was Robert von Ranke Graves he became the object of anti-German abuse from his jingoistic schoolfellows. 'My history', he writes, 'from the age of 14, when I went to Charterhouse, until just before the end of the war, when I began to think for myself, is a forced rejection of the German in me' (p. 30). The schizophrenic state of mind which his post-war dreams of battle induced in his opinion originates in the repression of vital areas of himself which the first of a series of British public institutions produced. Linked to his subjective experience of disintegration is another of the book's principal themes – the sickness of British society. Twenty-one of the thirty-two sections in the book are about his negative experience of Charterhouse and the British army.

Graves's portrait of his parents reveals another split, one which became a split in his own personality after he had internalized it. He associates his father – himself an establishment figure, an inspector of schools – with passing examinations, form-filling and a 'coldness . . . which is anti-sentimental to the point of insolence' (p. 16). Graves also debunks the second-rate verse his father wrote: 'That my father is a poet has, at least, saved me from any false reverence for poets' (p. 15). His mother, evidently the more powerful of the two parents, was also the more directly influential. He acknowledges how much he owes as a writer to her (p. 33) and the extent to which she, unlike the father, worried about him. But she was a religious puritan. In this capacity she represented another repressive instrument of the state, the protestant church. Her 'religious training', he writes, 'developed in me a great capacity for fear – I was perpetually tortured by the fear of hell – a superstitious conscience, and a sexual embarrassment from which I have found it very difficult to free myself' (p. 21). In his biography of Graves Martin Seymour-Smith points to a similar conflict in the young boy, one 'between maternally induced morality and independence'.[19] The extent to which his mother was responsible for his sexual gaucheness is born out early in the book when he recounts how terrified he was as a boy by any contact with young girls, outraged by any attempts on their part to make the mildest of sexual contacts with him. He writes with extreme candour of the continuance of his sexual inhibitions at public school and in the army. There is his long-lasting romantic attachment to 'Dick' (G. H. Johnstone) at Charterhouse beneath which lurked the potential bogy of sexual desire and fulfilment. There is his fastidious avoidance of prostitutes during the war. There is his mis-marriage to the feminist-minded Nancy Nicholson, the caricature scene of their wedding, the embarrassment of their honeymoon and Nancy's subsequent view of her marriage as 'a breach of faith with herself', leading to her wanting 'somehow to be dis-married' (p. 242). Graves is quite ruthless with himself in exposing the confusion in sexual identity that haunts his relations with both men and women. The consequence of this confusion is alienation from others and from himself. Everyone appears to be playing an absurd part in a universal farce.

The farce was played out on a grand scale with the outbreak of the Great War. For men at the front the only reality was their dangerous, exhausting and inhuman existence in the trenches, a hell-on-earth. This divided them from the population back in England which appeared to have lost touch with reality as the fighting men experienced it. Once again Graves found himself divided from his roots. Invalided back to England in 1916, he was appalled at the war fever everywhere: 'The civilians talked a foreign language, it was a newspaper language. I found serious conversation with my parents all but impossible' (p. 188). He proceeds to quote a mawkish letter published in the *Morning Post* by 'A Little Mother' offering up 'the human ammunition of "only sons" to fill in the gaps' (p. 189). The civilization for which he was fighting had succumbed to an insane public rhetoric. Accordingly Graves uses a grotesque rhetoric of his own in the autobiography to correspond to the general insanity around him at this time. It must have appeared particularly foolish to him in view of his German blood ties.

Graves's sense of alienation from his parents and the civilian population at large was further extended to the senior army staff responsible for much of the unnecessary slaughter. For him one of the worst features of army life was the presence of class distinctions in a situation where the only reality was the distinction between life and death. The autobiography traces the origins of his own class consciousness back to his middle-class home where at four and a half, as he puts it, 'I suddenly realized with my first shudder of gentility that two sorts of Christians existed – ourselves, and the lower classes' (p. 19). The 'gentility' of class is neatly undercut by his use of 'Christians' (instead of, say, 'humans') for whom class should be irrelevant. Home and school between them had made him a typical bourgeois snob of his day. On entering the army he found the same class stratification. It was the cause of numerous humiliations for him. It even prevented him from being posted to France because of his un-gentlemanly dress sense. His introduction to the Second Battalion of the Royal Welch Fusiliers is rendered as a caricature scene of the divisive effects that class distinctions had on men ostensibly fighting together against a common enemy. On entering the mess, Graves is first studiously ignored, then addressed as a 'wart' and humiliated in front of everyone present for a minute irregularity in his dress. Yet, at the same time that he hears with disbelief about the Battalion's ridiculous snob practices in the officers' mess, he feels compelled to record his recognition that it is a superior fighting unit to the more democratic regiment to which he had been attached before. The dichotomy within him stems from the divisive social structure from which he is struggling to detach himself sufficiently to become its critic. The end result is to turn him against the entire hierarchical society in which he has been brought up. But since he is also a product of that system he finds himself turning on himself. This is especially evident when, faced with Sassoon's protest in 1917, Graves could only offer a counsel of despair: 'I took the line that everyone was mad except ourselves and one or two others, and that no good could come of offering common sense to the insane. Our only possible course would be to keep on going until we got killed' (p. 215). Such was the extent to which Graves had become alienated from life within and outside himself by the third year of the war.

But his death wish is not fulfilled. At the end of the war he finds himself trying to accommodate himself to a society which has lost none of its insanity. The real problem lies in the fact that he carries within himself many of the elements against which he found himself in ostensible revolt. For example he has no moral qualms in pulling rank on a village policeman who had blocked his application to join the Special Constabulary. Because he is still an officer on the pensioned list Graves refuses to accept from him a third-class railway warrant – 'He and I might find ourselves in the same compartment,' he remarks ironically, 'and it would never do for us two to mix socially' (p. 281). But the object of his irony is as much himself as the policeman – both are warped by the sick society to which they belong. The self-destructive mood into which the war had driven him survived it and governed all his actions during the following decade. His marriage, his shop-keeping, his brief immersion in parish politics, his professorship in Egypt are all rendered in terms of high farce, because that is how Graves in his continuing state of alienation saw them. Since he was also participating in them he reserves his keenest satire for his own part in the proceedings.

His only solution, he concluded in 1929, was to say goodbye to it all in print and in person. The book, he wrote in 1929, 'is a story of what I was, not what I am'.[20] His aim in writing it was 'forgetfulness, because once all this has been settled in my mind and written down and published it need never be thought about again'.[21] It is hardly surprising that such total amnesia eluded him in his post-1929 existence. From the evidence he gives us in the autobiography one can understand why – because he had internalized so many of the conflicts dramatized in external form by the war. Continuing class consciousness is only one instance among many. For example, although he emerged from the Great War ostensibly opposed to war in general, he nevertheless volunteered – unsuccessfully in the event – for infantry service in 1939, as he casually admits in his 1957 Epilogue. But what *Goodbye to All That* did help him to achieve was a workable reintegration of his fragmented personality. It enabled him to make fun of those aspects in himself that were least acceptable without denying their continuing presence in him. As he writes at the end of the Epilogue, 'If condemned to relive those lost years I should probably behave again in very much the same way' (p. 282). To achieve that insight into himself Graves had to survive another war and come to the realization that art for him was a life-long means of maintaining his psychic equilibrium, initially in his autobiography and subsequently in his poetry.

Siegfried Sassoon: *The Complete Memoirs of George Sherston*

According to Siegfried Sassoon, the difference between his attitude to the Great War and that of Graves was that whereas Graves 'seemed to want the war to be even uglier than it really was', he

'wanted the war to be an impressive experience'.[22] That it failed to
live up to his expectations largely explains why it continued to
haunt him for the rest of his life. In 1964, three years before he died,
Sassoon was still trying to resolve the conflicts set up by his
experiences at the front: 'I was quite sure I should be killed (and
only by a hair's breadth wasn't) and all my diatribes against the war
ended in being recommended for a DSO! The whole thing now
seems incredible and mad, but it still has me in its grip.'[23] In
Sassoon's case the war helped to release him from a prolonged
adolescence (he was twenty-seven when he enlisted), for which he
was grateful, as well as subjecting him to the horrors of the trenches
which haunted him for the rest of his life. Little wonder that it held
him in its grip for so long. During those four years between 1914
and 1918 he lived life at a pitch which his earlier and later life never
reached. His literary reputation rests largely on the poems he wrote
during the latter part of the war and on the six volumes of
autobiography (four of which include the war years) that he spent
almost two decades of his later life writing.

According to Sassoon it was Edmund Gosse who planted in him
the germ of the idea out of which his first volume of autobiography
grew. After the Armistice celebrations of 1918 Gosse suggested to
Sassoon that he establish his literary reputation by writing a long
poem embodying typical country figures whom he had met in his
sporting days. At the time Sassoon rejected the idea as frivolous.
Ten years later Gosse was largely proved right when Sassoon
published *Memoirs of a Fox-Hunting Man* (1928) which overnight
brought him to the attention of a wider public. In this and the
subsequent two volumes of his memoir Sassoon hid behind the mask
of a fictionalized persona, George Sherston, and used pseudonyms
for every character except two. *Memoirs of a Fox-Hunting
Man* spans his life from the age of nine (1895) to almost thirty
(April 1916). For most of it he concentrates in a highly selective
way on his experiences of cricket and hunting. In the last portion of
the book he traces his career in the army from his idealistic
enlistment to his disillusionment with the entire war. *Memoirs of an
Infantry Officer* (1930) continues his account from April 1916 to
July 1917. This volume shows his transformation from a self-
destructive mock hero to the courageous if ineffectual author of his
famous letter of protest against the conduct of the war. *Sherston's*

Progress (1936) completes the trilogy; it covers his final year of fighting, from July 1917 to July 1918. The volume opens with his confinement to Craiglockhart mental hospital and follows his fortunes to his return to the Western Front where he was wounded by an inadvertent shot in the head from his own sergeant and invalided out of the army. In 1937 the trilogy was reissued as *The Complete Memoirs of George Sherston*. Sassoon then went on to write three volumes of straight autobiography: *The Old Century and Seven More Years* (1938) covers his first twenty-one years from 1886 to 1907; *The Weald of Youth* (1942) spans the years 1909 to 1914; and *Siegfried's Journey 1916–1920* (1945).

Where the last two volumes of straight autobiography overlap with the Sherston trilogy they concentrate on his literary life which Sassoon had denied to his alter ego, George Sherston. This exclusion has led several modern commentators to assert, in the words of Paul Fussell, that 'the *Memoirs* is in every way fictional'.[24] The publication of *Siegfried Sassoon Diaries 1915–1918* in 1983 does not confirm this assumption. In the first place not just the Sherston trilogy but *Siegfried's Journey* (which is purportedly non-fictional) make equally selective use of the diaries, omitting whole months at a time. Then there is the fact that both sets of autobiographies remain heavily dependent on the diaries. If anything, *Sherston's Progress*, the third part of which consists entirely of diary extracts over a four-month period, makes more faithful use of the diaries than does *Siegfried's Journey* written nine years later. In a revealing passage in *Siegfried's Journey* Sassoon refers the reader to *Sherston's Progress* to justify the omission from his straight autobiography of any account of his life during the first half of 1918. He says of Sherston: 'His experiences were mine, so I am spared the effort of describing them' (p. 69). He then proceeds to illustrate the difference between his two renderings of the same self: 'But Sherston was a simplified version of my "outdoor self". He was denied the complex advantage of being a soldier poet' (p. 69). This distinction does not turn Sherston into a fictional protagonist. It simply takes the process of autobiographical selection further than is usual for the genre. The simplification is probably responsible for the wider popularity of the Sherston memoirs, since George Sherston is someone with whom the majority of readers can empathize in a way that is not possible in the special case of a 'soldier poet'. But Sassoon paid a heavy price for his simplification of his earlier self. When he was coming to the end of the Sherston trilogy he realized, as he wrote in *Sherston's Progress*, 'in my zeal to construct these memoirs carefully, I have eliminated too many of my own self-contradictions'.[25] The autobiography may have been a success with the public but it didn't answer his more personal desire to understand better the conflicts which raged within him during the war years. But having exploited the material concerning his life as a soldier in the memoirs, when he came to write his straight autobiography he felt unable to repeat himself. The result was that his three later volumes of autobiography suffer from a similar one-sidedness and concentrate most of the time on the aspiring poet and writer. In *Siegfried's Journey* he claimed a different distinction between the two accounts of his life: 'Being fourteen years older than when I wrote Sherston's subjective account, I am now more inclined to analyse and investigate the inner history of a course of action.'[26] While this is true of the later trilogy it does not make of it the more interesting account as one might expect from this added generic element of self-reflection. Probably this is because he cannot apply this interpretative bent to his life as a soldier which evidently was the most powerful experience of his lifetime. It is as if he suffers the same dichotomy in his life and in the autobiographies in which he sought to resolve that dichotomy.

Even *Memoirs of a Fox-Hunting Man* makes frequent reference to what he calls his 'clarified retrospection'.[27] This first volume sets out to recapture his unthinking youth, 'that unforeseeing actuality which retrospection can transmute into a lucid and orderly emotion' (p. 179). Already the subjective experience recorded in his diary of the time is being refracted in the more objective narrative of the older writer, still anxious to recapture youth's immediacy of experience but aware of the need to make it intelligible through selection and interpretation. At one point Sassoon reproduces a brief entry in his diary recording a day's hunting with the Coshford staghounds in order to illustrate the difference between the perspectives of the diarist and autobiographer. Regretting the paucity of information in the diary, he explains: 'Since the object of these pages is to supply that deficiency I must make my reminiscent deductions as best I can' (p. 163). What follows is a denunciation of his shallow, inconsiderate youthful self. His earlier self concealed behind the image of a hard-bitten hunting man his poor horsemanship. He showed no pity for the terrified stag. He shook his whip at a parson shouting 'Brutes!' at the huntsmen. His analysis of the diary entry ends: 'All the sanguine guesswork of youth is there, and the silliness; all the novelty of being alive and impressed by the urgency of tremendous trivialities' (p. 165).

It is evident from *Siegfried's Journey* that early in 1914 Sassoon underwent an emotional crisis in which he felt that his 'life was being wasted on sport and minor poetry' (p. 53). *Memoirs of a Fox-Hunting Man* jumps over this period for purposes of contrast. Part Eight ends in July 1913 with Sherston broke and aimless after a season's hunting with the Packleston pack. Part Nine opens in September 1914, five weeks after he had joined the army with war already declared. The purpose of this literary juxtaposition is to contrast the pointlessness of his sporting existence with the 'reality' of life in the army. Yet this retrospective desire to impose a pattern on the past is in conflict with the actuality of that past. The diary version of his youth is that of 'sanguine guesswork', 'all the novelty of being alive'; it is the later reflective writer who concentrates on its 'silliness' and 'trivialities'. His crisis of 1914 did not alienate him overnight from his love of the countryside and of his favourite rural pastimes, hunting and cricket. The war diaries show that whenever home leave made it possible he would be off with the hounds without a thought for the pointlessness of it all. By then, of course, the 'reality' of war had revealed its own monstrous form of pointlessness. What has happened is that the later Sassoon has tried to impose on the earlier Sherston the soldier's temporary reactions at the front against the mindlessness of his country existence. But this is at odds with Sassoon's continuing love of the country and its pastimes during and after the Great War. The linguistic skill with which he has combined the two personas in a phrase like 'the urgency of tremendous trivialities' creates the illusion of a third integrated self such as he was seeking to realize in the autobiographical act.

Yet it is just this tendency to oversimplify the complexities of his personality that is responsible for Sassoon's long line of autobiographical works all of which represent attempts to rid himself of his obsession with the war. To read his diaries for the war period is to appreciate the extent to which Sassoon fluctuated between alternating moods and attitudes towards the war. He is forever adopting a posture which a new development forces him to abandon as inadequate, only to find himself back in the original belief some time later. His diary entries for 1915 equate war with an inner feeling of peace and freedom: 'I don't want to go back to the old inane life which always seemed like a prison. I want freedom, not comfort'.[28] By June 1916 he is beginning to sound less sure of himself, vacillating between his sense of freedom at the front and the sensation of safety he experiences whenever he returns home. He can only record: 'I'm in a desperate muddle tonight' (p. 75). To equate life at the front with freedom he had made light of the risk of death. But by the end of 1916 he realizes that 'the thought of death is horrible, where last year it was a noble and inevitable dream' (p. 109). Back in France in 1917 he loses the pacifist convictions he acquired in conversation with his Garsington friends back in England and becomes 'conscious of the same spirit that brought him serenely through it last year; the feeling of sacrifice' (p. 137). On his next leave he ricochets once more towards a pacifist stance and tries to make his public protest against the war. But once he is back in France the following year at his own request he congratulates himself for having escaped his own individuality by becoming once more 'the happy warrior type of officer' (p. 251). The month before his head wounds put him out of the war for good Sassoon is at his most confused, declaring in his company mess, 'I want to go up the Line and *fight*!', only to retire to his room where he writes 'on the monstrous cruelty of war and the horrors of the front line' (p. 269). No wonder he records in his diary the following day that he is 'beginning to realize the difficulties of combining the functions of soldier and poet', adding a footnote, 'soldier and pacifist' (p. 270).

The diary records Sassoon's mounting confusion as he tries unsuccessfully to reject the country life he loves and embrace life in the trenches. Both the rejection and the embrace are falsifications of a more complex reality. Country existence does not have to be a mindless one. There are more ways of experiencing love, sorrow and hate ('things which every poet *should* know!', he remarks in his diary, (p. 52)) than through participating in the senseless slaughter of the Great War. Besides, life in the trenches can be as mindless as fox-hunting, while even the world of fox-hunting is 'a clearly defined world, which is an idea most of us cling to', he writes in *Memoirs of a Fox-Hunting Man* (p. 208).

Sassoon's problems arose from the simplification of his past self that he projected in the character of George Sherston. Much of his motivation for the courageous, even foolhardy way in which he conducted himself at the front came from his desire, as he wrote in his diary, to 'let people see that poets can fight as well as anybody else' (p. 53). More realistically he saw the experience of fighting as material for his poetry: 'Great thing is to get as many sensations as possible' (p. 51). These poetic considerations, denied to his outdoor self, George Sherston, are crucial to understanding the crisis he reached in 1917 culminating in his protest and – even more incomprehensible in the memoirs – his later desire to return to the front. For his presence at the front, he recorded in his diary, as well as supplying material for his poetry, won respect for the anti-war sentiments he expressed in his poems. If his protest could do no good then perhaps his exposure of the futility of war in his poetry might prove more effective.

Yet even this explanation is too simple. The fact is that Sassoon was suffering from nervous exhaustion caused not simply by his length of service at the front but by the conflicts raging within him. As he writes in *Siegfried's Journey*, had there been any chance of his being passed fit for active service in 1917 he might well have forgone his protest in favour of a return to France 'in a spirit of self-destrucive bravado' (p. 56). That self-destructive streak in him points to a general tendency towards masochism that underlies much of his seemingly irrational behaviour. He recounts in *Memoirs of an Infantry Officer* how, struggling to formulate his protest, 'there was no relaxation of my inmost resolve, since I was in the throes of a species of conversion which made the prospect of persecution stimulating and almost enjoyable' (p. 202). The same leaning towards self-punishment surfaces in *Sherston's Progress* when he returns to active service after his spell at Craiglockhart hospital:

To feel in some sort of way heroic – that was the only means I could desire for 'carrying on'. Hence, when I arrived at Clitherland, my tragedian soul was all ready to start back for the trenches with a sublime gesture of self-sacrifice . . . It wanted to see itself . . . dying defiantly in some lime-lit shell hole; 'martyred because he could not save mankind', as his platoon sergeant remarked afterwards, in a burst of blank-verse eloquence of which he had hitherto believed himself incapable (p. 96).

Even the irony employed at the expense of his younger self shows the continuing presence eighteen years later of his pervasive masochism.

If he is forever punishing himself, from where does the guilt come? I would suggest it originates in what Fussell calls his 'binary vision', his habit of creating dichotomies in his life which force him to adopt extreme positions. Because he knows that these dichotomies falsify the reality he is haunted by guilt. Fussell argues that in exploiting these dichotomies in the memoirs, 'Sassoon is being critical of his art as well as of his wartime character.'[29] This seems to fly in the face of the evidence. Sassoon's binary structure in the memoirs, which becomes increasingly complicated with each volume, rather demonstrates the persistence of his binary vision into later life and explains why the composition of all six volumes of autobiography failed to rid him of his obsession with war. The simplest of all his dichotomies is that between the peace of rural England and the discord of war-torn France. *Memoirs of a Fox-Hunting Man* is structured on this oversimplified contrast. *Memoirs of an Infantry Officer* juxtaposes his two conflicting attitudes to the war, that of heroic self-sacrifice and that of outraged protest. Again this dichotomy proves unsatisfactory, because the first attitude was complicated by the anti-war poems arising from his self-immolation, while the second attitude ran counter to all the values he thought he was defending.

The difficulties this polarization causes Sassoon can be illustrated by the episode in *Memoirs of an Infantry Officer* in which he is sent to Lord and Lady Asterisk's country manor-house. His period of convalescence with the real Lord and Lady Brassey at Chapelwood Manor, Nutley, in Sussex is also described in his diary entries for May 1917 where, unusually, he records his conversation with the spiritualist Lady Brassey in the third person. It is as if he were detached from the character playing his part in a ridiculous conversation in which she maintains that 'death is nothing' and can find no excuse for his keeping out of danger. Naturally he is alienated by this particularly outspoken representative of the civilian population from which all soldiers at the front felt cut off. But, as Sherston admits, she 'symbolized the patrician privileges for whose preservation I had chucked bombs at Germans' (p. 187). He is caught in his own binary trap. As John Lucas has pointed out of this episode, Sassoon's 'rebellion was against all he had chosen to believe in and identify with. In short, it was against himself.'[30]

Sassoon's internal conflicts proliferate in *Sherston's Progress*. In this third volume he leaves the 'unreality' of Craiglockhart hospital for the 'reality' of war. His reasons for returning to the front multiply – it's the only way out; it offers forgetfulness; soldiers' company is better than civilians'; and so on. Unsure whether his suicidal wish to return to the war is a product of spiritual pride or war-weariness, he is uncomfortably reminded that he could choose personal freedom instead with the connivance of his consultant there. He is temporarily saved from these mounting conflicts within himself by a spell in the Middle East. But on his return to the Western Front he is plunged into a worse quagmire of self-contradictory attitudes and opinions. Although his latest justification for his presence at the front is to look after the men in his company, he goes out on a fateful night patrol 'to escape from the worry and responsibility of being a company commander' (p. 259). It is interesting that the older Sassoon at this climactic point in the book adopts the use of the third person for his former self, only to betray the same set of polarities still fighting within him: 'I should admire that vanished self of mine more if he had avoided taking needless risks. I blame him for doing his utmost to prevent my being here to write about him. But on the other hand I am grateful to him for giving me something to write about' (pp. 259–60). Here is the identical dilemma that he noted in his diary was confronting him back in 1915, twenty years earlier, that between heroic self-sacrifice and the desire to seek out material for his art.

Sherston's Progress shows Sassoon straining against the artificial structure he had established in the first volume of the memoirs. More than once he refers ironically to his 'neatly contrived little narrative' which is constantly about to 'come sprawling out of its frame' (p. 18). Clearly he realizes that the oversimplification that he has employed for artistic reasons has undermined not just his therapeutic aims in writing down his memoirs of the war but also the structural coherence of the third volume. This explains why he felt it necessary to write three more volumes of straight autobiography in which he hoped that the full complexity of his earlier life might be laid bare. It is not just a matter of reintroducing his poetic self. In place of the fictitious bumbling Aunt Evelyn, for example, in *Siegfried's Journey* he can reveal the fact that his mother's black-and-white view of the war was becoming a barrier between them (p. 27). *Siegfried's Journey* is also more critical than the memoirs were of the fact that the diary writer's 'behaviour was still unconnected with any self-knowledge' (p. 41), an acquisition which Sassoon implies should be more evident in this last volume of autobiography.

But having concentrated on one set of dichotomies within his personality in the Sherston memoirs, Sassoon found that he couldn't repeat himself. So the new dichotomies which he displays for the first time in the three volumes of straight autobiography once again present a partial picture of the complex totality. As if he realized this, he began with *The Old Century and Seven More Years* to indulge an escapist bent in his nature, what he calls his 'queer craving to revisit the past and give the modern world the slip'.[31] By the end of *Siegfried's Journey* he has abandoned all hope of mastering the dilemmas of his past by ordering them into the aesthetic framework of an impersonated or a straight autobiography. On the last page he poses a final unresolved dilemma surrounding 'the discrepancy between the art of autobiography and the rudimentariness of reality': 'Can it be', he asks, 'that the immediacy of our existence amounts to little more than animality, and that our ordered understanding of it is only assembled through afterthought and retrospection?' (p. 224). But his afterthought produced no more ordered understanding of the internal conflicts that had plagued him throughout the war than did his diary writing at the time. Twelve years later he became a convert to Roman Catholicism and chose to forget his wartime diatribes against Christianity. Perhaps religious confession achieved what literary confession had failed to do.

John Osborne: *A Better Class of Person*

The first volume of John Osborne's autobiography, *A Better Class of Person* (1981), which covers the first twenty-seven years of his life from 1929 to 1956, was not ostensibly undertaken for therapeutic reasons. According to Osborne, it was born, like

Graves's, out of penury: 'It was simply venal: *The Sunday Times* said they would serialize them.'[32] This is understandable when one bears in mind the fact that his last play, *Watch It Come Down*, was in 1976 and that *A Better Class of Person* took him three years to write. Nevertheless it so clearly confronts him with the childhood origins of his choleric adult personality (which hitherto mostly had to find indirect expression through the medium of his plays) that it is instructive to approach it as an attempt, however unconscious, at self-analysis. To have written his autobiography at this stage of his career has a kind of inevitability about it when one remembers to what extent all his later plays (after *Inadmissible Evidence*) largely abandon internal dramatic tension in favour of addressing the audience – or rather rebuking it – through mouthpieces like Laurie in *The Hotel in Amsterdam*, Pamela in *Time Present* or Wyatt in *West of Suez*. He claims that he wrote his autobiography entirely from memory. While the interpolated letters and extracts from notebooks qualify this claim, it still suggests a frame of mind anxious to uncover the roots of his present impasse.

The nature of that impasse appears to lie in the degeneration of an initially healthy determination to avoid the emotional sterility of the Osbornes and the Groves (his mother's family). '*Don't be afraid of being emotional,*' runs an entry from his Notebook of 1955. '*You won't die of it.*'[33] But the 26-year-old's determination to cultivate the habit of expressing his emotions at all costs turned over the years into the middle-aged playwright's penchant for denouncing all and sundry in his plays, for writing letters like his notorious 'I hate you, Britain' letter published in *Tribune* in 1961, or for denouncing the contemporary management of the Royal Court Theatre in terms so libellous that legal action was avoided only by his foreshortening the autobiography by three years and thereby omitting the offending passage.[34] The vitriol has been accompanied by a streak of cruelty as well, which surfaces in the autobiography in the at times vicious portrayal of his mother, still alive when it was published. Asked what if she were to read the book, Osborne replied: 'She hasn't the interest. Her self-absorption is total.' Then he added egotistically: 'Perhaps she would just be flattered at being there.'[35] The obsessive nature of his mounting literary offensive against the rest of the world, and the absence of any substantial positives in his work, give off a strong flavour of neurotic defensiveness.

The autobiography allows him to come face to face with the causes of his obsessions and offers him an opportunity for placating the ghosts from his past. Chief among these are his parents and their families. No one is better than Osborne at evoking the soulless atmosphere of lower middle-class or upper working-class suburban family life. The sheer detail with which he recollects the routine of his paternal grandparents' daily existence shows how early in life he became a detached observer of these awful family scenes. He describes, for instance, how the grandfather was treated as a servant by his indolent domineering wife who would not even let him sit down or have a cigarette before lunch. The couple are described with striking verbal wit, but underlying and often undercutting the linguistic dexterity is a feeling of rage and violence which belongs to the child as protagonist but seems infantile in the adult narrator. When, for example, Osborne complains that 'throughout my childhood no adult ever addressed a question to me' (p. 15), I recognize the exaggeration of the neglected young boy but feel uncomfortable that the 52-year-old successful writer should still repeat this myth with such conviction.

Doubts proliferate when it comes to Osborne's portrait of his parents. His ailing father slowly dying of tuberculosis can do no wrong, while his mother (whom he normally refers to by her forenames, Nellie Beatrice, or simply as 'Black Looks') is pictured as a monster of unfeeling banality. He dwells on her 'stillborn spontaneity' (p. 19), and 'her view of affection or friendship as a system of rewards, blackmail, calculation and aggrandizement.' 'Nothing strikes me with such despair and disbelief', he continues of her, 'as the truly cold heart' (p. 63). He can be extremely funny at her expense. There is a long, viciously comic description of her appearance which includes features such as her lips ('a scarlet-black sliver covered in some sticky slime named Tahiti or Tattoo') or her face which she covered with a 'powder called Tokalon, which she dabbed all over so that it almost showered off in little avalanches when she leant forward over her food' (p. 35). But the humour is invariably combined with an animosity which appears to be out of all proportion to the charges he makes against her. Her worst crime of all is to inform her son when she insists on taking him into the room where his dead father is laid out: 'Of course, this room's got to be fumigated, you know that, don't you? Fumigated.' Osborne observes: 'For the first time I felt the fatality of hatred' (p. 101). Quite apart from the obvious fact that his mother was in a state of shock at the time she said this, it is the continuation of that hatred into the present that jars the reader of this book so strongly.

What Osborne cannot forgive his mother for is her surviving his father who died when Osborne was twelve and therefore old enough to remember him. His mother, according to Osborne, appears to have taken her cue from his father's unloving mother who never forgave her son for getting asthma as a boy on a trip abroad and running up doctors' bills for several hundred pounds. Osborne identified strongly with the victimized father and built him into an idealized paragon, what he called back in 1969 'a man of tremendous probity and integrity'.[36] The facts that he supplies concerning his father do not substantiate this claim. The father spent much of his time away from home leaving his son to the small mercies of his harridan-like mother – if the son's description of her is to be believed. But Osborne, as Jeremy Treglown has pointed out, is similar to most of his leading characters who 'are preoccupied like Hamlet with their father's deaths'.[37] In the autobiography Osborne describes a repertory company's production of *Hamlet* which he staged and in which he played the part of Hamlet after having cut most of the other characters' lines. His temporary position as actor-manager allowed him to indulge himself in the performance of a lifetime, manhandling a supposedly lesbian Gertrude as lewdly as he could because, he explains, 'she had a residual taste of the rank sweat of Claudius's enseemed bed and of all the monstrous regiment of women'. He talks of 'the low, lunging coarseness of the Osborne Prince, . . . full of black looks rather than nighted colour', and remembers feeling a 'huge euphoria' that he 'was certain never to be allowed again' (p. 233). In fact he has gone on allowing himself this satisfaction ever since, a modern Hamlet using his mother's weapons ('black looks') to revenge himself on her and all womankind. Jeremy Treglown suggests that the loss of his father was responsible for his development of a personality that he has made 'heterosexual to the point of misogyny'.[38] Some of the acts of revenge on women which he recounts in the book (such as slipping a used contraceptive into one hated woman's smoked-salmon sandwich) tend to substantiate this charge.

However the autobiography and its subtext offers us occasional glimpses of the female prototype in his life without her son's myopic distortions. Several times he acknowledges the financial help she gave him during his penurious years in repertory theatre. For six or seven years in his twenties he spent long weekends with his mother for the sake of the free food. Yet all he can do is to grumble about their Black Mondays together when he was made to pay the price by accompanying her on her day off. As if to prove his point Osborne reprints a letter from this clearly lonely ageing woman that starts: 'I live here on my own: think well of those I love so dearly and it helps to make my life worth while' (p. 207). When she is allowed to speak in her own words Black Looks turns out to be just another fallible but loving mother. Another letter from her provides a glimpse of a more outward-going individual than Osborne depicts: 'I like life and people . . . Please *don't* think I am discontented: *far from it*' (p. 89). The evidence is sparse but sufficient to show the extent to which Osborne has excluded the more lovable half of his mother's character and directed all his undoubted verbal wit at her more dislikeable traits. Does she, then, embody for him all the potentialities in his own personality against which, mainly in dramatized personae, he has spent his life fighting?

Yet Osborne's rejection of the mother and all she represents is less successful than he assumes. As Jung suggests, 'Children are driven unconsciously in a direction that is intended to compensate for everything that was left unfulfilled in the lives of their parents.'[39] Near the end of the autobiography Osborne quotes a particularly revealing extract concerning his mother from his Notebook of 1955: ' "I'll say that for him – he's never been *ashamed* of me . . ." I am ashamed of her as part of myself that can't be cast out, my own conflict, the disease I suffer and have inherited, what I *am* and never could be whole' (p. 271). Here for a moment is the insight that is wanting in the majority of the book, the realization that all his diatribes against the world at large are the articulation of his mother's unvoiced grievances against the world which she thinks 'owes everything to her' (p. 206). It is a terrifying realization because it internalizes the conflict with his mother, a conflict which otherwise he can hold outside his psyche, as it were. The realization, however, belongs to Osborne as a young man, belongs in fact to the year in which he wrote *Look Back in Anger*. It is a realization that largely eludes the ageing author of the autobiography.

A *Better Class of Person* contains passages as effective as any in his best plays belonging to the earlier part of his career. But it suffers, like the later plays, from the predominance of the bad mother he has imaged for himself. Like the Nellie Beatrice of his imagination, he has had to find someone outside himself to blame for all his misfortunes. Osborne's autobiography exemplifies a common use of the genre to reinforce an existing neurosis rather than to trace it back to its origins in the early self and thereby to see its inapplicability to his adult needs. A failure in self-analysis does not necessarily spell failure for the autobiography in which it is embedded. Oversimplification of the complexities of personality, as in the case of Sassoon's memoirs, may defeat the therapeutic aims in undertaking an autobiography but can positively add to its readability. Even then the writer's dissatisfaction with the distortions and evasions that ensue is likely to surface in the text in some form. But self-illusion on the grand scale, such as one meets in Osborne's autobiography, leads the reader to withhold the absolution which the writer of a confession implicitly seeks. For one thing Osborne does not apply to himself the same satirical treatment that he meets out to his mother and to most of the other women in the book. Then his vitriolic use of language to discharge emotions more proper to the child than to an author in his fifties similarly convinces the reader that he is being asked to condone a version of events which is clearly a distortion even of the facts that he is given. This author is more concerned to add to his neurotic defences than to explore openly the origins and nature of his personality.

'Personality,' Jung wrote, 'as the complete realization of our whole being, is an unattainable ideal. But unattainability is no argument against the ideal, for ideals are only signposts, never the goal.'[40] The confessional mode, it appears, commits the writer to a higher degree of self-exposure than do other modes of subjective autobiography, and the greatest of its practitioners, Augustine and Rousseau, have both recognized this necessity where Osborne has sought to substitute in its place sarcasm, irony and wit. Because the humour operates at the expense of the confessional urge Osborne's use of the genre serves only to expose the unconscious evasions, distortions and omissions which enable him to transform his earlier self into the triumphant if mock-heroic hero of his autobiography.

8

Myths and Dreams in Autobiography

Dreams have always fascinated men by virtue of their enigmatic and fragmented communications. They hold out the tantalizing hope that a different life or a different meaning to life awaits the dreamer who can break their code and read the messages of his dream self. Where in earlier times dreams, especially those of a premonitory nature, tended to be appropriated by religion or the state, the Romantics turned to them for individual self-enlightenment. For De Quincey in particular the dream mechanism is what he calls 'the magnificent apparatus which forces the infinite into the chambers of the human brain, and throws dark reflections from eternities below all life upon the mirrors of that mysterious *camera obscura* – the sleeping mind'.[1] Dreams have their own language, he maintains, a private language which he spent a lifetime trying to decipher and share with his readers. Man's understanding of dreams underwent its most celebrated revolution with the publication of Freud's *The Interpretation of Dreams* in 1900. In it he claimed that the 'interpretation of dreams is the royal road to a knowledge of the unconscious activities of the mind'[2] and asserted that a 'dream is a (disguised) fulfilment of a (suppressed or repressed) wish'.[3] The symbolism employed in dreams is, according to Freud, also commonly found in popular myths and legends. He saw myths as 'distorted vestiges of the wishful phantasies of whole nations, the *secular dreams* of youthful humanity'.[4] Both myths and dreams, then, can give covert expression to man's repressed desires. They offer glimpses, however indirect, of a more primitive aspect of the self which can achieve social acceptance only in such disguise. Buried within their obscure powerful images lie incest, murder and a whole range of feelings that have become taboo in the course of man's evolution.

Since Freud's time the potent presence and influence of the unconscious in us all has become something of a twentieth-century preoccupation. One might therefore have expected a greater number of autobiographers of this century to have given a more central role to dreams in their autobiographies than is in fact the case. Yet Freud himself illustrated how difficult it is to translate and make public the language of one's dreams. In the 1935 Postscript to his own disappointingly reticent autobiography he refers the reader to *The Interpretation of Dreams* for information about him of a more personal kind. Yet, as Michael Sprinker has shown, when he turns to his celebrated dream of 'Irma's Injection' he breaks off his interpretation in alarm at mid-point. The pseudo-autobiographical text, therefore, 'enacts', Sprinker suggests, 'the very process of repression that it has sought to illuminate'.[5] It isn't simply that Freud is afraid of revealing too much of his carefully guarded private self (though he is). It is that dreams always contain, as Freud puts it, 'a tangle of dream-thoughts which cannot be unravelled'.[6] So that an autobiographer seeking to tell his dream life is faced by his own inability to interpret the dream language beyond a certain point. He has two choices. Either he can report and comment on his dreams, in which case the report is likely to do an injustice to the power of the original while the comment is liable to be misleading because of the resistance of the tangle at the core of all dreams if Freud is to be believed. Or he can choose to follow the later De Quincey who, especially in *Suspiria de Profundis* prefers to enact his dreams without comment so as to reproduce the peculiar language and structure of the subconscious mind without intervention by the waking self. Of course the waking self cannot help from interpreting in the course of transferring the dream into a written narrative. But the more successful he is at rendering the dream unmediated the greater the danger that its meaning will become impenetrable.

The prototypal modern autobiography in which dreams and the inner life of the unconscious occupy the major portion of the book is Jung's *Memories, Dreams, Reflections* (1962), undertaken (in collaboration with Aniela Jaffé) in his eighty-third year. Jung's autobiography tells the story of his internal life to the near-exclusion of his external life. He sees his life as 'a story of the self-realization of the unconscious', so that the only events in his life worth telling 'are those when the imperishable world irrupted into this transitory one'.[7] This, he explains, is why he speaks chiefly of inner experiences among which he includes his dreams and visions. As a child of three or four he felt he was first initiated into the dark secrets of the earth through an archetypal dream he had of a subterranean phallus enthroned like a god. 'My intellectual life had its unconscious beginnings at that time,'[8] he asserts. It is remarkable how often Jung found himself resolving major cruxes in his life through the medium of dreams in which his unconscious found more direct expression than during his waking life. Two dreams directed him as an indecisive school-leaver to take up medicine as a career. A series of fantasies and dreams in 1914 helped him to abandon the heroic idealism which was impeding his growth as an analyst and which on an international scale he held responsible for the outbreak of world war and to confront his unconscious – an essential step in his search for self-understanding. A further series of dreams led to his discovery of the 'anima' within him. It was also via a dream that he came to understand how his journey to North Africa was in reality a search for his primitive shadow self which living in Europe had repressed. Most of his important dreams in this way helped put him into connection with the larger life of the race and constitute a potent metaphorical means of overcoming the egotism inherent in the autobiographical act.

Jung explains at the beginning of his autobiography that he intends to tell what he calls his 'personal myth' or 'fable'. At least one British autobiographer has been directly influenced by Jung's book. This is Christopher Isherwood who described *Christopher and His Kind* (1976) as 'an attempt, influenced by Jung, to explore one's personal sexual mythology and identify one's sexual archetypes'.[9] Throughout his life Isherwood found himself compellingly attracted to a particular type of boy. In the book he pursues this obsession back to its origins. His archetypal love object is a young boy who is both working class and foreign. This archetypal stranger-lover embodies for him the opposite of all he was in revolt against – his mother's expectations that he would get married, the middle-class ethos into which he was educated, and the imperialism and militarism of his native country. The material of his life is organized within the text around this central myth in which he cast himself and his friends in appropriate roles. 'My book is rather internal,' Isherwood wrote, 'and when I describe people I will try to concentrate on their myths, the myths I made about them.'[10] These myths have already been incorporated in his fiction and his later inability to disentangle myth from actuality is circumvented by concentrating on distilling what they meant to him in his private mythology. It is interesting to note the distinction he makes between *Lions and Shadows* with its reticences and *Christopher and His Kind* which he hails as true autobiography in view of the fact that in both books friends like Spender and Auden receive the same mythological treatment and are made to play similar mythic roles which he has assigned to them in the carefully scripted drama of his life.

Dreams and the mythopoeic imagination tend to coincide in the structural role which archetypes play in autobiographies of this kind. It is not that such autobiographers exclude external events, but they subsume them into a larger pattern which traces the shape of their subconscious life. In both Jung's and Isherwood's autobiographies, for example, their sojourns abroad are imbued with a symbolic significance which directs the reader's attention to the role they play in the inner life of each of them. Thus Isherwood, for example, from the start of his book sounds a recurrent note whenever he encounters anything associated with America. He is anticipating the finale where he leaves Europe and the dissatisfied self that he associates with his exile on the Continent and sails for America which is destined to be (as he knows) his new homeland. It is also destined to be the country in which he meets the Swami who transforms his life and helps him rediscover his underlying religious nature. So in describing the ten years leading up to his permanent departure for the States, Isherwood is contriving to describe the workings of his unconscious as it leads him back to the inner core of his true self which he had vainly tried to deny throughout that period.

In a similar way Forrest Reid's *Apostate* (1926) and Edwin Muir's *An Autobiography* (1954) both give priority to the inner self. They achieve this focus not by excluding the facts concerned with their personal histories but by subordinating them to their myths and dreams the importance of which they signify by making of them structural motifs within the text. In both cases the inner life also connects them to the collective life of the race and so offers them a form of religious satisfaction without which external life seems to each without meaning or purpose.

Forrest Reid: *Apostate*

Forrest Reid is a currently neglected novelist who spent all his life from 1875 to 1947 in Ulster apart from three years at Cambridge. All of Reid's books take for their subject boyhood and its world for which he had a life-long attraction. Boys were an obsession with him and his books about them were, according to him, 'mere pretexts for the author to live again through the years of his boyhood'.[11] When he was just turning 50 he finally took the natural step of writing directly about his own childhood in *Apostate*, an autobiography of his first twenty years. On its appearance he declared that it would never have a continuation because it would be impossible to treat his later inner life with the same degree of frankness while to keep it 'external and objective . . . would not really be a sequel'.[12] In fact he did eventually bring out a further autobiographical volume, *Private Road* (1940), a compilation that collects his reflections on his own books, other writers, Cambridge and issues of particular interest to him such as a chapter on dreams. But it is not, like *Apostate*, a work of imaginative literature nor an account of his inner experiences.

Born in Belfast, Reid was the youngest son of a family of six children. His mother came from an ancient aristocratic line and felt brought down in the world when her husband's business went bankrupt. His father died in 1881 when Reid was in his fifth year. A few months after his father's death his childhood nurse, Emma Holmes, whom in the absence of a truly loving mother he adored, left suddenly, a traumatic event in his young life which he recalls vividly in *Apostate* as marking the end of his childhood. After her departure he found compensation in a private dream world which originated in a serialized story about some children's quest for heaven which Emma had read to him. His dream was of an idealized landscape in which he played with a boy companion guarded against intrusion from the outside world by mythological stone figures. Eventually at the age of sixteen or seventeen the dream stopped recurring. Having failed to find an intimate friend during his boyhood he met at work a fellow apprentice, Andrew

Rutherford, with whom he fell in love. *Apostate* ends with a description of the moment in 1896 when Reid cemented his new friendship by showing Andrew his private journal in which his innermost life and predilection for boys was revealed. Reid's discovery to Andrew Rutherford was to cause an eventual rupture in their friendship, something which *Apostate* avoids mentioning by closing where it does.

Forrest Reid was a pederast, though not a practising one as far as can be ascertained. His idealized love for younger boys doomed him to perpetual disappointment. He could never fully express his love and the boys inevitably grew into men. All forms of homosexuality according to Freud are narcissistic, since the homosexual directs his love towards those of his own sex who show a similarity to his image of himself. In the case of a pederast that image is of his own boyish self. Reid's cult of boyhood appears to be an attempt to recover the intense happiness of his first six years with his nurse Emma, first through the medium of his recurrent dream, then through fictionalized evocations of it in his novels and in *Apostate*. Reid's life ended, in effect, when his dream stopped recurring in late adolescence. After that his external life was a shadow compared to the substantiality of his childhood memories and the fantasies he constructed from them. 'I could get on swimmingly', he wrote in *Private Road*, 'until I reached my King Charles's head – the point where a boy becomes a man.' 'Then', he continues, 'something seemed to happen, my inspiration was cut off, my interest flagged . . . I supposed it must be some form of arrested development.'[13] The only way left for him to recapture the regal delights of his boyhood was by endlessly writing about it, sublimating his repressed sexuality into works of refined art.

The obsessional nature of his subject matter virtually predetermined Reid's view of art. *Apostate* opens with a definition of art which subordinates external events to a visionary view of both life and art:

The primary impulse of the artist springs, I fancy, from discontent, and his art is a kind of crying for Elysium . . . an Eden from which each one of us is exiled. Strangely different these paradisian visions . . . only in no case, I think, is it our free creation. It is a country whose image was stamped upon our soul before we opened our eyes on earth, and all our life is little more than a trying to get back there, our art than a mapping of its mountains and streams (p. 3).

Once again one finds the childhood autobiographer evoking the archetypal image of Eden to describe the joys of his earliest years. Reid does not attempt to deny the validity of the external life in order to assert the substantiality of his dream life: 'There are two worlds, and it never occurred to me to ask myself whether one were less real than the other . . . I lived in both' (p. 72). *Apostate* spends as much time describing the external as the internal life because it concentrates on those years when first of all his external life was blissfully happy, then on the period in which he internalized that feeling of bliss in the form of his archetypal dream, and finally on his late teens when the dream deserted him but he thought for a short time that he had found his dream companion in the external world. Nevertheless the unseen world controls the shape of the narrative and reflects the adult author's values and preferences. Inevitably he was accused of escapism, a charge he accepted but cleverly redefined in a letter in which, he wrote, he 'preferred the literature of escape, and what *I* should call the literature of imagination for the escape is only from the impermanent into the permanent'.[14]

The consequence of this adult preference for the inner world over the outer world which held equal sway in his childhood is to create a tension in the autobiography between his conscious and unconscious attitudes towards the genre he was using. On the one hand he has described the ease with which he wrote *Apostate*: 'I had nothing to change, nothing to invent, nothing to ponder. I simply watched and listened while the whole thing was re-enacted before me.'[15] This accords with his approach to fiction which, he wrote to Clemence Dane, should avoid any form of intervention by the narrator: 'Everybody has a King Charles's head, and mine is a terror of the didactic, of the slightest intrusion of the author as a commentator, teacher, or chorus.'[16] On the other hand Reid's attempt to let his earlier self take command of the autobiography, leaving to his fifty-year-old self the task of the faithful recorder, is sabotaged in the opening pages by the same conscious adult self that finds it cannot bow out of the text it is producing. Instances of such intrusion occur when he presents his adult theory of art, or when he acknowledges the sheer impossibility of reproducing his earlier self as it was before it became overlaid with later memories as well as lacunae in his adult recollections of childhood:

It is difficult . . . to understand why, when for once the historian sets out deliberately on a voyage of rediscovery, the way should suddenly reveal itself as beset with pitfalls, and even the prime actor in the tale should have taken fright and withdrawn himself into a hiding-place. It is as if we actually could get closer to truth through fiction than through fact (p. 5).

The fact is that the prime actor is as reliant on the author for his lines as any fictional character. Reid goes on to draw an analogy between looking back at his earlier self and peering aged five or six through the coloured glass door of the conservatory in the Botanic Gardens through which he saw, not palms and cactus, but an entire tropical landscape of his imagination. All too soon he was dragged away into the public gardens in the light of a dull October morning: 'Looking back through Time is very much the same as looking through that greenhouse door. The shapes of things remain unaltered, but there is a soft colour floating about them that did not exist in the clear white light of morning' (p. 7). For a moment he shows an awareness of the extent to which his nature inclines him to idealize early boyhood. This realization leads him to attempt a reconciliation between past and present selves. He continues: 'Only, again, I am not sure – am not sure, I mean, that this clear white light ever did exist for me. I cannot help thinking that I was in those days very much what I am now' (p. 7). Far from separating his adult from his childhood self, Reid here suggests an identity between them. In effect he is claiming that his distortion of the past to create his private dream world occurred the moment that the present became the past back in his childhood. What he doesn't seem to be willing to recognize is that the same distorting mechanism continued to operate on his already distorted recollection of his early life, both inner and outer. His private dream world, then, is the product of a lifetime's continuous distortion. The act of literary reconstruction actually constitutes one more stage in the transformation of actuality to dream and myth.

This epistemological confusion on Reid's part stems from his life-long desire to recapture the unspoilt joy of infancy. Even the joy is a fiction, for it was Emma who dragged him away from the conservatory door. Letters which his sister, Fanny, wrote to him show that he has exaggerated the extent of his mother's neglect

('Mother adored you,' she claims) and of his isolation within the family after Emma left ('After she went you were petted and spoilt by everyone.').[17] Yet it is important to the fifty-year-old author to write as if he were able to reproduce untarnished his childhood vision of the world around him: 'That Emma precisely resembled the Emma I now see when thinking of her is more than doubtful. I was a boy of six when she left us, and a child's vision of grown-up people is true perhaps at any time only for other children. As I recall her she is ageless . . .' (p. 16). As Brian Taylor suggests, *Apostate* actually 'is as much an account of the dependency of the grown man on the memory of boyhood as of the circumstances of that boyhood itself'.[18] Reid's very dependency is what causes him to conceal from himself his adult presence in the book. Only that way can he recapture the miracle of being a boy. He confirms this supposition in *Private Road* when he talks about the experience of writing one episode in *Apostate* in which the landscape reveals its mystic spirit to him: 'While I was writing these words the whole incident seemed to happen over again.'[19] This need to focus exclusively on his past self in order to re-experience in the present the joys granted to that earlier self also accounts for the mesmeric quality of the prose. This insinuatingly draws the reader into Reid's trance-like state in which the world of dreams and the outer world intermingle without any sense of incompatibility.

This phantasmagoric quality in his writing carries the reader across the pauses between sentences and often paragraphs and inevitably shows at its strongest when he is describing his private dream world. His composite reconstruction of the recurrent dream landscape of his boyhood can be illustrated only by a longer quotation. His first full-length description of it comes at the end of Part Two which contains his account of the desolation he felt in the period following Emma's departure:

It was always summer, always a little after noon, and always the sun was shining.

The place was a kind of garden . . . Always when I first awakened I was in broad sunlight, on a low grassy hill that was no more than a gentle incline, sloping down to the shore. A summer sea stretched out below me, blue and calm . . . I saw no house or building, no sign of human habitation. But I did not feel lonely. I knew there were people there, and that I could find them if I went in search of them . . .

I was waiting for someone who had never failed me – my friend in this place, who was infinitely dearer to me than any friend I had on earth. And presently out from the leafy shadow he bounded into the sunlight. I saw him standing for a moment, his naked body the colour of pale amber against the dark background – a boy of about my own age, with eager parted lips and bright eyes. But he was more beautiful than anything else in the whole world, or in my imagination.

. . . It was as if I had come home; as if I were, after a long absence among strangers, once more among my own people. But the deepest well of happiness sprang from a sense of perfect communion with another being. It was this I looked forward to, this that I still longed for when I awoke . . .

There were animals, large and small, wild and tame, but all perfectly friendly – even the great jewelled snakes, coloured in fire, who would lift up their flat watchful heads from the long grass and gaze at me through the sunlight. These beasts were my playmates, too . . .

In some of these glades were stone figures watching over our playground, and guarding it from intrusion. A spirit was alive within them: they could hear me and see me, though no breath passed their lips (pp. 72–5).

Reid goes on to describe one particular crouching beast in black marble, with a panther's body and human face, which had something terrible about it: 'I had the feeling that if it were to rise up and open its mouth and send forth its cry across the sea, this would be the end of all things' (p. 76). It is a vividly realized dream world that Reid contrives to imbue with his own spellbound sense of wonder. It is also remarkably revealing about the nature of the dreamer. The origins of his ideal companion are suggested by a later passage in *Apostate* describing another dream in which he is bending over a pool:

. . . when I leaned down I had looked straight into the eyes of a dark and almost swarthy face, which, as I remembered now, bore no resemblance to my own. And yet it had caused me *then* no surprise, it had been the image I expected to see . . . All I could decide was that I very much preferred this dream shape, and would change my own for it even now, and at the risk of not being able to establish my identity (pp. 102–3).

Isn't his ideal companion the shadow self he sees reflected in the pond? 'In each of us', Jung writes, 'there is another whom we do not know. He speaks to us in dreams and tells us how differently he sees us from the way we see ourselves.'[20] Reid's dream is a perfect example of wish–fulfilment, perpetuating the Edenic state of his infancy when Emma allowed him to play by himself, guarding him like one of those stone beasts from outside interference. The dream also suggests a narcissistic form of homoeroticism for which he found confirmation when eventually he turned his back on Christianity to cultivate the animistic beliefs of Hellenistic culture. But when he first began having this dream it is natural that most of its features should derive from the Garden of Eden (both Emma and his mother being devout Protestants). That image of primal innocence has enabled so many childhood autobiographers to associate their own memories of a pre-Lapsarian infancy with the archetypal myth of the origins of Western civilization, as I showed in Chapter 5.

But what about that terrifying image of the marble beast which threatens his dream world with extinction? More nearly awake than the other stone beasts it nevertheless feels like cold marble to the touch. In my opinion it clearly resembles the sphinx, half lion, half woman, which terrorized Thebes. By correctly answering its riddle Oedipus rid Thebes of this pest and was rewarded with the throne and the hand of his mother in marriage. But what if the sphinx and the mother are one and the same person? In *Apostate* Reid blames his mother for failing to understand his private world as Emma had done. He says of her: 'It was as if, deep down below the surface, something within me, reaching out tentatively, was met by a blank wall of insensibility' (p. 38), and he goes on to talk of her 'power to inflict a wound' that 'did not heal' and 'is there still' (pp. 38–9). His timid attempts to contact his mother, like his hesitant contact with the cold hard surface of the marble beast, are countered by his fear of the power that he might unleash if he brought either of them fully to life. Better to restrict himself to the confines of his private dream world where he is king than risk its exposure and destruction by admitting primal woman into that world.

So Reid remained fixated in his autoerotic dream life, always looking for its realization in the external world, yet knowing that 'no earthly love could ever fill its place' and that its memory was 'like a Fata Morgana', leading him 'wherever some faint reflection of it seemed for a moment to shine' (p. 74). Why, then, does he end *Apostate* by describing the scene in which he shares his private life with Andrew Rutherford, leaving the reader with the impression that here at last he has found his dream companion in the external world? His journal entry for that night and the letter Andrew Rutherford wrote him immediately after that shared evening together make it clear that at least for a brief moment Reid had every reason for thinking that his dream had come true.[21] But that was not possible because of the self-enclosing nature of the dream. Reid's only means of perpetuating his dream and its momentary (if mistaken) realization is through his art. The act of writing about the secret life of his boyhood and ending the story where he did simultaneously recaptured its magic and perpetuated it in the unchanging shape of his autobiography. Mary Bryan suggests that 'if a friend could not understand the interiority of Reid's conflict, then the anonymous reader who does understand becomes the perfect friend who shares the journey to awareness'.[22]

But did Reid need anyone else beside himself? Doesn't he make plain in the book his preference for the company of his dream self? The reader is allowed to observe at a distance Reid communing narcissistically with himself, trying desperately to lose his adult consciousness in contemplation of the child who never wanted to grow up. Ironically it is fortunate that Reid fails to eliminate his adult presence. That presence is responsible for the organization of the narrative around his central dream scenario and it possesses the linguistic skill to evoke his inner landscape with power and beauty. His very desire to exclude his older self contributes to the reader's understanding of the author-narrator's personality. The book actually enacts Reid's failure as an autobiographer to escape into the world of his childhood, a failure which characterized the reiterative pattern of Reid's adult life. What he found in the course of writing the book was that autobiography, unlike fiction, will not allow the writer to avoid 'the intrusion of the author as commentator, teacher or chorus'. Out of the tension this produced came an autobiography in which inner vision explodes into the outer world but in which equally the representative of the outer world, the narrator, is forced to interpret and give meaning to the inner vision of the protagonist.

Edwin Muir: *An Autobiography*

Dreams play an equally central role in Edwin Muir's classic autobiography where they represent less wish–fulfilments, as is the case with Reid, than what Jung calls 'a spontaneous self-portrayal, in symbolic form, of the actual situation in the unconscious'.[23] For Muir one's life can be described either externally in terms of its 'story' or internally in terms of the 'fable' it aspires to. The facts which comprise the story of his life appear to him not just external but deceptive, offering 'as it were a dry legend which I had made up in collusion with mankind'. Here he appears to dismiss external accounts of the self as a product of social conditioning. Like Jung Muir believes that 'the life of every man is an endlessly repeated performance of the life of man'.[24] What puts individual man in touch with his species are above all dreams which offer him 'glints of immortality' (p. 54), fragments of the fable that never fully reveals itself. It is for this reason, he explains, that 'no autobiography can confine itself to conscious life, and that sleep, in which we pass a third of our existence, is a mode of experience, and our dreams a part of reality' (p. 49). When he was in his early thirties Muir underwent two years of analysis. This helped release in him a flood of dreams and waking visions which enabled him to recover his childhood sense of wholeness. When the analysis was broken off he claimed to have attained a more objective view of himself through the agency of his dreams:

I saw that my lot was the human lot, that when I faced my own
unvarnished likeness I was among all men and women, all of whom had the
same desires and thoughts, the same failures and frustrations, the same
acknowledged hatred of themselves and others, the same hidden shames
and griefs, and that if they confronted these things they could win a certain
liberation from them (p. 158).

The progression in this sentence from 'I' to 'they' both represents his own progress from a neurotic obsession with himself to a sense of his shared humanity and indicates the presence of a didactic strain in the book whereby, like a religious adept, he offers his own path of enlightenment as a source of inspiration to his readers.

Muir's early life had been sharply divided between an idyllic childhood spent on his father's various farms on the Orkney Islands and his first five years in Glasgow to which the family moved when he was fourteen and where both parents and two brothers all died within a short period. 'I was too young for so much death,' he reflects (p. 104). 'I climbed out of these years, but for a long time I did not dare look back into them' (p. 110). At twenty-five he went to work for two years in a bone factory in Greenock, a horrific experience that he could not bear to retell even to his wife, who first learned about it when she read his autobiography.[25] He got married in 1919 and moved to London where he worked as an assistant to Orage on *The New Age* to which he was already contributing poems and aphorisms. After two years there (during which he underwent his analysis) he left with his wife for Prague.

This is the chronological point at which Muir's first venture into autobiography, *The Story and the Fable* (1940), breaks off. Written shortly after Muir had turned fifty, it ends with 'Extracts from a Diary, 1937–1939', the years during which the book was gestating. In the interval he had wandered around Europe with his wife earning his living as a professional writer and translator. Eventually he settled in St Andrews where he wrote the book. A diary entry for April 1938 shows that the autobiography evolved out of a renewed need to explore himself and his past at a time when he was suffering from severe depression:

The problem: to discover what I am; to find out my relation to other people. The first an inward problem, the second an outward one. For when I understand myself I will have changed myself; and when I understand my relation to other people, I will have set about changing society . . . I must take stock of my own life, and though that is in appearance an investigation of the past, I know it is important for the future: because of this change that it must bring about.[26]

The writing of *The Story and the Fable* was a substitute repetition of the analytical process he had undergone two decades earlier. In the spring of 1939 while he was in the middle of writing the autobiography, he experienced a reconversion to Christianity. His rediscovery of religious belief confirmed his long-held conviction in man's immortality, a view which is a leading motif in the book. But he was not prepared to subscribe to any particular sect or church (calling himself 'a sort of illicit Christian')[27] and therefore avoided introducing his specifically Christian beliefs into *The Story and the Fable*. After completing the book at the end of 1939 he confessed to a friend that he still had not formulated a philosophy of life for himself:

I find with some dismay, after going over my life, that I have no philosophy . . . I believe that I am immortal, certainly, but that in a way makes it more difficult to interpret *this* life . . . I have no explanation, none whatever, of Time except as an unofficial part of Eternity – no historical explanation of human life, for the problem of evil seems insoluble to me: I can only accept it as a mystery . . .[28]

By 1952 *The Story and the Fable* was out of print. To help sell a reprint his publishers urged him to update it. Muir agreed, but found the task more difficult than he had anticipated. During the intervening decade he had entered more directly into public life by working for the British Council in Edinburgh, Prague and Rome. To reflect this alteration in life-style in the autobiography meant changing his approach from one of retrospective contemplation to a more narrative one. His inner life during the intervening period had been mainly a matter of his growing religious convictions, but he was reluctant to make these the central theme of the continuation. In the end he focused his narrative on the rise of impersonal authority in modern Europe that he had witnessed at first hand before and after the Second World War:

The picture I am trying to present is of power growing more and more impersonal as it becomes more mechanically perfect, and of the greater part of mankind, its victims, who contrive, in the greater and greater stresses, to remain human. Omnipotent impersonality on the one side, fumbling and unequipped humanity on the other.[29]

P. H. Butter has pointed out the ostensible connections between the two parts of the enlarged autobiography, published as *An Autobiography* (1954). Where *The Story and the Fable* (equivalent to the first six chapters of *An Autobiography*) treats the loss and recovery of his childhood experience of immortality in personal terms, the continuation dwells on the way the loss of that experience in so many has made possible the growth of impersonal power. Opposed to impersonal power is the imagination which fosters personal communication and which is itself reinforced by the arts.[30] Another critic, Elgin Mellown, has offered a different interpretation of the way in which Muir unifies the disparate material of the enlarged autobiography. He sees the revised book as 'a circle growing ever larger, the author at the centre being aware first of only his own feelings and ideas, but moving then toward the perimeter of man in all his relationships – to nature, to society, and to God'.[31] Nevertheless it seems undeniable that whereas in the first half the story was continuously interpreted by reference to the unfolding fable, in the second half the story all but eliminates the presence of the fable. Since I am focusing here on the use of dreams and myth in autobiography I will confine my remarks largely to the first six chapters of *An Autobiography* which are a lightly revised version of *The Story and the Fable* now inaccessible to most readers.

Muir felt that in *The Story and the Fable* he had finally found a form that suited him. Yet the search for self-knowledge was his overriding concern, even if this entailed sacrificing form and other aesthetic requirements to it:

I don't know whether I want artistic unity; I would like to avoid all make-believe; the arranged patterns of modern novels give me such a stale, second-hand, false and tired feeling. And I have no wish to confess either, for the sake of confession: I am too old for that: I want some knowledge; it really comes to that.[32]

This urgent quest for self-knowledge that inspires the autobiography originated in his period of analysis almost two decades before he wrote the book. But, as the ending of *The Story and the Fable* shows, the search continued throughout the relatively leisurely intervening period: 'Without those few years of idleness, of looking and looking back, I might never have really lived my life.' The process of discovery was two-fold. First he re-experienced his past life, especially his prolonged childhood. Then his present life 'came alive too as that new life passed into it; for it was new though old'.[33]

So an understanding of childhood is of key importance to Muir in coming to terms with one's adult life. After he had completed *The Story and the Fable* he wrote an article, 'Yesterday's Mirror: Afterthoughts to an Autobiography', in which he distinguished three ways of seeing life all of which are 'related in some way to childhood, positively or negatively'. The three ways, which are also described in his poem of this time, 'The Three Mirrors', are those of the realist who has forgotten his childhood and sees only the triumph of evil, of the mature man who has remembered his childhood and 'sees in the mirror an indefeasible rightness beneath the wrongness of things', and of the greatest poets and mystics at their greatest moments who share with the child a vision of the world 'in which both good and evil have their place legitimately' and all opposites are reconciled.[34] Visionaries like St Augustine and Blake journey back as well as forward through life to rediscover the innocence of childhood. The poem suggests that Muir himself spent his life up to the time of his analysis in turning his back on his childhood convinced that the world was 'askew'. Between then and the writing of *The Story and the Fable* he reverted to his past and his outlook changed to the second way of seeing life in which 'locked in love and grief/Good with evil lay'.[35] The third stanza, starting 'If I looked in the third glass', emphasizes with that conditional opening the fact that the visionary reconciliation of opposites belongs to a part of the fable from which he is denied. As he explains in the article, 'in life we are ourselves the opposites and must act as best we can'. Expelled from paradise, Muir is bound to the life of conflict which is the lot of fallen man.

But he can think back to his childhood and recognize in it some elements of the universal fable. The Orkney he was born into, he writes, 'was a place where there was no great distinction between the ordinary and the fabulous' (p. 14). Originally his parents were to him 'fixed allegorical figures in a timeless landscape' (p. 24) whose changelessness made them more solid, 'as if the image "mother" meant more than "woman" and the image "father" more than "man" ' (pp. 24–5). One could say that childhood gave him the only experience of immortality possible to mortal man, since for a child alone (apart from the mystic and the poet) time does not exist. 'Certain dreams convince me that a child has this vision, in which there is a completer harmony of all things with each other than he will ever know again' (p. 33). Soon enough the vision is broken by the introduction of death and guilt into his life. But dreams offer the grown man a new means of access to the child's experience of immortality without which, Muir feels, life seen as mere animal existence is unendurable. In fact, he argues, the impulse to know himself and his belief in immortality are connected by the importance to both of them of the unconscious self revealed in dreams and waking visions. During sleep, Muir asserts, 'there seems to me to be knowledge of my real self and simultaneously knowledge of immortality' (p. 54). Obviously for him the real self is immortal.

Accordingly Muir gives considerable space in *The Story and the Fable* to descriptions of his more significant dreams, which he insists are part of our total experience, perhaps the part that reminds us most of our connection to the fable. By 'fable' Muir means what Jung calls 'the collective unconscious' to which Muir makes explicit reference in one of his published letters and which clearly caught his imagination at the time of his analysis.[36] Twice he refers in the autobiography to the 'racial unconscious' (pp. 57, 164) and it is clear from the book that he attached greatest significance to those dreams of his which were 'ancestral' or 'Millennial' (p. 56). In this kind of dream his own experience, for example, of witnessing the killing of farm animals, is related to the ancient world, which instinctually felt a connection between men and animals and into which 'our unconscious life goes back' (p. 47). This understanding of the way dreams put us in touch with our racial unconscious is virtually identical to that which Jung held. As Jung put it: 'All consciousness separates; but in dreams we put on the likeness of that more universal, truer, more eternal man dwelling in the darkness of primordial night. There he is still the whole, and the whole is in him, indistinguishable from nature and bare of all egohood.'[37]

Muir offers a perfect example of the way personal memories and experience from childhood impinge on racial memory and myth by describing a dream he had in which he is led by a Christ-like figure into a hillocked landscape where he sees all the great animals with their heads raised heavenwards. The dream incorporates images of his childhood landscape in the Orkneys, of his childish picture of Christ and of his childish understanding of what his parents meant by the Millennium. But the dream also connects those personal images and memories with man's immemorial sense of guilt at his need to kill animals for his survival. 'All guilt seeks expiation and the end of guilt, and our blood-guiltiness towards the animals tries to find release in visions of a day when man and the beasts will live in friendship and the lion will lie down with the lamb' (pp. 54–5). His dream, he claims, is about the origins of man and his sense of unavoidable guilt as much as it is about the Millennium when men and animals can become finally reconciled. It is about beginnings and ends, about the three mysteries that Muir feels possess our minds: 'where we came from, where we are going, and . . . how we should live with one another' (p. 56). In dreams the child and visionary combine to produce images of harmony that belong to the fable.

'I should like to write that fable,' Muir says, 'but I cannot even live it' (p. 49). All he can do is to recognize certain stages in the fable –

'the age of innocence and the Fall' (p. 49). The predominance of
Christian imagery and symbolism in most of Muir's dreams is
understandable in the light of his religious upbringing in a spirit of
Revivalist Christianity. He skilfully employs this segment of biblical
myth to structure his autobiography. His use of this much repeated
metaphor of childhood differs from the more romanticized use to
which many other childhood autobiographers have put it in that he
sees the whole of his life up to the time of writing about it as a
personal experience of the myth. His Fall and Expulsion come not all
at once but extended over the second half of his childhood. First at
the age of 7 came a sense of fear and guilt without any apparent
cause. Muir's inability to talk to his parents about feelings he did not
understand first brought home to him his growing separation from
figures who had before seemed to him omnipotent. The same year he
had his first experience of school, a fallen state in which he found
his earlier freedom curtailed. For Muir the fruit of the Tree of
Knowledge had a particularly bitter taste. In later life he had a
nightmare vision of 'an enormous school . . . and millions of
children all over the world creeping towards it and disappearing into
it' (p. 42).

But it was not until he was 14 when the family moved to
Glasgow that his expulsion from the paradise of his childhood was
complete. The horrors of his first five years in Glasgow during
which time both parents and two brothers died confirmed his sense
of final Expulsion. In the autobiography he contrasts the good life
of Orkney where 'their life was an order, and a good order' with
Glasgow where they 'were plunged out of order into chaos'(p. 63).
The principle underlying Glaswegian society was 'competition, not
co-operation, as it had been in Orkney' (p. 93). This is the nature of
his fall and of modern man's fall since the Industrial Revolution. In
an extract from his diary for 1938 printed at the end of *The Story
and the Fable* (but not in *An Autobiography*) he gives vivid
expression to this modern version of the Fall: 'I was born before the
Industrial Revolution and am now about two hundred years old.'
On his arrival in Glasgow he 'found that it was not 1751, but 1901,
and that a hundred and fifty years had been burned up in two days'
journey'.[38] Like Adam, Muir sees the world outside paradise as
cursed ground from which in sorrow he is forced to make a living
for the rest of his life. Nothing is more symbolic of the industrial

wasteland into which he had been expelled than his two years' experience of the Fairport bone factory with its pervading smell of decay and burning maggots and his permanent sense of physical shame: 'We lived in Fairport in a state of chronic reprobation, always in the wrong, among the filth and stench, grinding out the profits' (p. 133). What could be more reminiscent of God's judgement on Adam: 'In the sweat of thy face shalt thou eat bread, till thou return unto the ground; for out of it wast thou taken.'[39]

Redemption first came to Muir in the form of his analysis and the flood of dreams this released in him. Out of this experience came what he describes as 'a conviction of sin, but even more a realization of Original Sin' (p. 158). His dreams also included one of universal purification, a kind of mass baptism in which even 'the harlots of France' are washed clean in a river that sweeps away the distinctions between good and bad which characterize the first two ways of seeing life in his article 'Yesterday's Mirror'. Dreams also restore to him his childhood sense of immortality, the presence of the Tree of Life in the Garden of Eden. He recounts one waking dream in particular in which he dives off a star back to an icebound earth. Realizing that he would be shattered by the impact, his soul shot out of his body and watched the body smash into the ice where it is eaten up by a huge kind of walrus. The soul continues to hover until out of the hole in the ice into which the walrus retreated emerges himself reborn while under his feet new grass springs up. In that one dream alone he has moved from the biblical myth of creation and the fall to apocalyptic visions of death, destruction and resurrection into a state akin to paradise regained.

After he experienced a series of waking visions of which this was one he began to realize that his Nietzschean cult of personality in his later Glasgow years was hopelessly timebound. He even has a dream of Nietzsche self-crucified out of pride. This reaction against Nietzsche led him to conclude that 'immortality is not an idea or a belief, but a state of being in which man keeps alive in himself his perception of that boundless union and freedom, which he can faintly apprehend in time, though its consummation lies beyond time' (p. 170). So his dreams lead to his conviction that 'human life is not fulfilled in our world, but reaches through all eternity' (p. 170). This central belief in man's immortality has important repercussions for his understanding of the true nature of the self, the subject of his autobiography. 'A personality', he concludes after his period of analysis, 'is too obviously a result of a collaboration between its owner and time, too clearly *made*' (p. 181). The adult persona is nothing but 'a crude imitation of (the child's) romantic conception of some grown-up figure' (p. 67), a childish make-believe. The only answer is 'not to cultivate but to get rid of personality' (p. 181), which is precisely what he tries to do in his self-portrait by giving priority to the interior over the exterior life. The argument is circular, since dreams form the core of the interior life and they constitute the evidence for his belief in immortality.

The problem in giving dreams a central place in an autobiography is one of interpretation. Their powerful emotive content is liable to unbalance the reader, arousing expectations of significance that remain unfulfilled because of their private and indirect nature. Muir overcomes this difficulty by recounting those dreams of his which establish a connection between the personal circumstances of his life and the recurrent patterns of human life in general. Almost every dream he describes reinforces his belief that 'the life of every man is an endlessly repeated performance of the life of man' (p. 49). His dreams vividly enact the essential drama of his life and 'go without a hitch into the fable' (p. 115). In this way he is able to describe the minutiae of his own life while connecting it to archetypal patterns of human behaviour. This allows the reader the dual satisfaction of being able to observe and participate in Muir's dream world. His dream of his mother's death is also a dream about the immortality of the soul. His other Glasgow dream of the fight between a big man and a little man who wears out his much stronger opponent prefigures, I would suggest, the way in which the childhood self insists on asserting itself despite the repressions of the neurotic adult self. His various dreams in which he brings back to life inanimate figures connect with man's persistent belief in rebirth, both psychological and spiritual.

Muir wrote in a diary extract reproduced in *The Story and the Fable*: 'The Eternal Man has possessed me during most of the time that I have been writing my Autobiography.'[40] It accounts for much of his selection of material and for the way in which he structures and interprets it. The fable of Eternal Man has offered him a blueprint for his own version of the fable which constitutes one of the major modern contributions to the genre.

9

The Spiritual History of the Self

In looking at specifically twentieth-century writers' autobiographical accounts of the self it is fitting that I should have resorted to the terminology of analytical psychology which has revolutionized our conception of human nature in the course of the century. Yet what today we refer to as the 'self' or the 'psyche' might be thought to be continuous with what was called the 'soul' for centuries before the process of secularization set in with the nineteenth century. According to Shumaker it is probable that the rise of modern subjective autobiography coincided in England with the kind of religious introspection encouraged by Calvinist theology.[1] Many of the most outstanding early examples of introspective autobiography are religious accounts of the soul's struggle with despair. Because the Protestant ethic made each human being responsible for his own spiritual life the autobiographies of men like Bunyan and Fox appear to be describing the same internal conflicts and search for wholeness that twentieth-century autobiographers describe in secular terminology. Just as Bunyan's spiritual struggle can be viewed, as J. N. Morris suggests, as 'an account of a man's struggle with his neurotic constitution',[2] so one can argue that Muir's neurosis and analysis bear a remarkable resemblance to the spiritual aridity followed by religious conversion which is the central ingredient of most seventeenth-century non-conformist autobiographies. It will be remembered that Muir set out to write his autobiography as a renewed attempt at analysis, or rather self-analysis, at a time when he was suffering from severe depression. It seemed quite natural to him that in writing about his past for therapeutic purposes he should discover a renewed belief in Christianity, and the vocabulary he employs in *The Story and the Fable* is equally indebted to psychoanalytical literature and the Bible.

The few overtly religious autobiographies of this century that have acquired a reputation invariably equate religious with psychological states of mind. Thus G. K. Chesterton in his *Autobiography* (1936) prefaces the chapter dealing with what he calls his 'period of madness' with his genuine belief that he has known the Devil in person.[3] Similarly C. S. Lewis in *Surprised by Joy* (1955) sees his childhood nightmares (the product of inadequate ego defences according to Ernst Jones) as constituting 'a window opening on what is hardly less than Hell'.[4] He also describes the stage at which he converted his introverted practice of self-examination into what he calls the extrovert practice of prayer: 'I had been, as they say, "taken out of myself".'[5] Such parallelism points to an underlying similarity between these ways of confronting the self. The models of such experience and the vocabulary that accompanies them change with time or culture or belief. To some extent words like 'soul' and 'psyche' are constituted by the methods used for identifying and describing them. But Bunyan and Muir, Rousseau and Leiris would all probably accept Jung's assertion that 'the spiritual adventure of our time is the exposure of human consciousness to the undefined and indefinable'.[6] The quest is the same. What differs is the method of approach and the terminology employed. It is perhaps easier for someone like Bunyan who has at his disposal an entire vocabulary widely understood by his contemporaries to write of this common adventure than it is for G. K. Chesterton or C. S. Lewis who can no longer assume such an understanding. It is even more difficult for the twentieth-century autobiographer who wishes to describe his spiritual history without recourse to the closed systems of Christianity or analytical psychology. W. B. Yeats immediately comes to mind. Brought up in an age of growing disbelief by his domineering atheist father, he spent the rest of his life searching for a satisfying outlet for his spiritual nature. His *Autobiographies*, which themselves constitute stages in his search, record the desperate measures he felt forced to adopt to replace Christianity: 'I am very religious', he writes in *The Trembling of the Veil*, 'and deprived by Huxley and Tyndall, whom I detested, of the simple-minded religion of my childhood, I had made a new religion, almost an infallible Church of poetic tradition, of a fardel of stories, and of personages, and of emotions, inseparable from their first expression, passed on from generation to generation by poets and painters with some help from philosophers and theologians.'[7] Yeats's lifelong spiritual quest, charted in *Autobiographies*, encompasses not just the world of literature; it also takes him on incursions into spiritual seances, the Irish nationalist movement, Indian mysticism, and Gaelic peasant culture. His urgent need to fulfil his religious yearnings led him to shape his life to answer this need, and further to shape his autobiography to reflect this lifelong search for spiritual fulfilment. This is why the book is so episodic. But each episode in his life is given significance by interpreting it as a fragment of the divine life.

It is only by the exercise of his considerable mythopoeic powers that Yeats contrives to give spiritual significance to his fragmented life. He is suffering from a disadvantage that, according to Roy Pascal, afflicts all autobiographers after the classic period of autobiography stretching between Rousseau (1782) and Goethe (1831). During that brief half-century, Pascal argues, there was a feeling of trust and confidence in the spiritual wholeness of the self and the meaningfulness of its destiny that disappears from later autobiography.[8] Nevertheless it is a feeling that some modern autobiographers seek to recapture in the course of their lives without recourse to shared systems of established belief. Rather than trying to reanimate the timeworn concepts of Christianity by borrowing from the dramatis personae of psychoanalysis in the manner of Chesterton, such autobiographers have preferred to model themselves more on Yeats's example, forging their lives into mythic patterns that offer satisfaction to their spiritual yearnings. In this chapter I have chosen two very different autobiographies by W. H. Hudson and John Cowper Powys to illustrate the way in which the spiritual history of the self is a far more difficult because more individualistic undertaking for a twentieth-century writer than it was for Christian autobiographers from Augustine to Newman. Both Hudson and Powys, like Yeats, felt compelled to turn away from Christianity in spite of their strong religious natures. Each is forced to seek within himself the sources for his renewed belief in the spiritual significance of his own life and of life in general. At the same time, like Reid and Muir, they both give priority to an inner life which helps them to understand their external history and to give shape and meaning to it in their autobiographies.

W. H. Hudson: *Far Away and Long Ago*

The circumstances leading to the composition of Hudson's childhood autobiography bear a resemblance to the kind of vision
sometimes granted to religious mystics at a crucial turning point in
their lives. At the age of seventy-four Hudson was suffering from one
of his recurrent bouts of severe illness that periodically afflicted him
as a result of the rheumatic fever which had nearly killed him when
he was sixteen. Recovering in a convent nursing home at Hayle in
Cornwall during November and December 1915, he suddenly had
what seemed like total recall of the childhood he had spent in
Argentina. The 'vision' continued and during his six weeks'
confinement in bed he wrote a long first draft of the autobiography.
Aware of the possibility that he might die of his illness, he was struck
all the more forcefully by the effect of these memories of his happy
childhood which appear to have acted as a form of self-cure, just as
visions frequently precede religious conversion in seventeenth-
century autobiographies. Throughout the next three years of the war
he revised and shortened his original draft which in its longer version
was too formless for his liking. Published October 1918, *Far Away
and Long Ago: A History of My Early Life* was immediately greeted
as a masterpiece in its genre.

Prior to his moment of recall Hudson had been reading with
great admiration Aksakoff's *Childhood* to which he twice alludes in
his own autobiography. In his opinion Aksakoff was quite
exceptional in that, unlike most people, he was able to re-create the
feelings of childhood uncontaminated by the distortions of an
adult's memory, so that 'in his case the picture was not falsified'.[9]
Hudson claimed that he also had never outlived his childhood
experience, especially of nature, so that his autobiography likewise
relives the past 'in its true, fresh, original colours' (p. 196). Hudson
has a strong personal motive for wishing to believe that his
childhood memories flood back to him unmediated by an adult
consciousness. It is because he has a Romantic's worship of the
child's proximity to the marvels of the natural world which he
wants to retain into later life. At the same time he is well aware of
the generic need to select, arrange and shape this material, even if
he refers to it in distinctly negative terms. Normally, he writes,
'unconscious artistry will steal or sneak in to . . . falsify the picture'

(p. 195). The fact is that he felt compelled to 'reshape' his first draft. He also justifies the hiatus of three years over which he skips prior to the last three chapters by explaining the need 'to bring [the book] to a proper ending' (p. 248). His references in the book to the classic childhood autobiographies of St Augustine, Leigh Hunt and Aksakoff in themselves point to Hudson's consciousness of the generic tradition to which he was contributing. So he uses all the guile of an autobiographer to create the impression of a guileless recorder of his past.

The first twenty-one chapters of the autobiography concentrate on Hudson's memories of childhood between the ages of three and twelve (1844–53). The remaining three chapters dwell on his fifteenth and sixteenth years with another jump in the last chapter to his eighteenth year when his mother died. His parents had emigrated from New England to a farm on the Argentinian plains near Buenos Aires in the 1830s. The fourth of seven children, Hudson enjoyed unusual freedom as a child, roaming at will through the pampas by foot or on horseback with little supervision. He never went to school, and received only spasmodic private tuition. Much of the book is taken up with descriptions of his wanderings through the country-side, his observation of the animals and plants he meets there and of the neighbouring settlers whom he gets to know. The coda consisting of the last three chapters recounts his two successive illnesses from typhus and rheumatic fever from which the doctors forecast that he would never recover, his mother's death, and his gradual loss of religious faith which gave place to a personalized form of animism that he still clung to in old age.

Far Away and Long Ago has a rather awkward, idiosyncratic shape to it. But the autobiography is held together by the older Hudson's unusual sense of continuity with his childhood self and especially with his childish sensitivity to and joy in the feelings that nature in the wild aroused in him. The opening chapter offers a sample of the sort of experiences from which he constructs most of the book. There is the description of the farmhouse where he was born with its distinctive row of twenty-five Ombú trees followed by a disquisition on the nature of these and other trees native to the region. There are also loving evocations of his dog, his mother, and two grotesque occasional visitors to the house. We appear to be reading the childhood account of an unusually perceptive naturalist. But in what way is it a spiritual history of the self? Hudson is careful to prepare for his later introduction of those animistic beliefs which became for him a religion in adult life. He does this by emphasizing his unusual capacity as an adult to recapture his childhood feelings with all their original strength when in the presence of nature. Recalling the stream running in the plain behind the house, he writes:

Whenever we went down to play on the banks, the fresh penetrating scent of the moist earth had a strangely exhilarating effect, making us wild with joy. I am able now to recall these sensations, and believe that the sense of smell, which seems to diminish as we grow older . . . is nearly as keen in little children as in the inferior animals (p. 6).

His autobiography confirms him in his belief that he has lost none of his childhood responses to the natural world. In 1893 Hudson published *Idle Days in Patagonia*, a semi-autobiographical collection of essays based on his sojourn there during 1870–1. In it he expresses most openly his conviction that in early childhood we come closest to the emotions experienced by our primitive ancestors in a state of nature. Citing Thoreau's description of his exuberant desire to seize a woodchuck stealing across his path and eat it raw, Hudson comments: 'In almost all cases . . . the return to an instinctive or primitive state of mind is accompanied by this feeling of elation, which, in the very young, rises to an intense gladness, and sometimes makes them mad with joy, like animals newly escaped from captivity.'[10] Hudson's description of his childhood in his autobiography attempts to recapture this sense of an animal's unthinking enjoyment of an uninhibited existence in the wild. Compared to the numerous chapters devoted to his adventures in the pampas, the few descriptions he offers us of his mother and the family home make one realize that he felt his first home was out of doors. He says as much in *Idle Days*:

The greenness of the earth; forest and river and hill; the blue haze and distant horizon; shadows of clouds sweeping over the sun-flushed landscape – to see it all is like returning to a home, which is more truly home than any habitation we know. The cry of the wild bird pierces us to the heart; we have never heard that cry before, and it is more familiar to us than our mother's voice.[11]

It is interesting that Hudson is more explicit about these most intimate feelings in *Idle Days* than in *Far Away and Long Ago* where they act as an implicit thematic element that helps hold the book together. I would argue that the autobiography is structured on the gradual replacement of home and parents by the world of nature.

Opposed to nature in the wild stands civilization. For Hudson 'the civilized life is one of continual repression'.[12] This is how his spasmodic experiences of schooling are received in the book. Trigg, his first tutor, seems to the young boy's eyes a Jekyll and Hyde figure, loved by the adults but a tyrant in the schoolroom. 'For', Hudson explains, 'he was a schoolmaster who hated and despised teaching as much as children in the wild hated to be taught' (p. 23). Trigg, in fact, is as oppressed by the civilization whose values he is attempting to transmit as are Hudson and his brothers. Civilization is given its most potent symbolic expression in Hudson's description of Buenos Aires, the capital of a country torn by civil wars which seem to erupt from the city and spread like plagues across the plains until they exhaust themselves in bouts of pointless bloodletting. Even his first visit there in his sixth year left him shocked by the behaviour of the idle young gentlemen who baited the laundresses and nightwatchmen for a pastime and the savage beggars that infested the streets. But his later stay there when he was fourteen provides the excuse for a diatribe against the evils of civilization in general. The city he remembers is contaminated by the stench of rotting cattle flesh from the slaughtering grounds at its southern end. He compares these open abattoirs to the worst scenes in Dante's *Inferno*, pervaded by 'the smell of carrion, of putrefying flesh, and of that old and ever-newly moistened crust of dust and coagulated blood' (p. 250). Nowhere else in the book is his language quite so violent as in this description of Buenos Aires as 'the chief pestilential city of the globe' (p. 251). As a child of nature he inevitably falls victim to this city of infection by catching typhus. His early conviction that 'our loss in departing from nature exceeds our gain'[13] receives personal confirmation in the near-fatal illness he succumbs to.

In the major portion of the autobiography chapter after chapter vividly evokes the plant and animal life that Hudson spent much of his childhood observing and entering into. Repeatedly trees or birds are treated as fellow beings with a life of their own that is quite as thrilling to the young boy as the human fauna he encounters on the plain. Even the trees assert their individuality as Hudson wittily demonstrates in the case of a rare acacia: 'Of all our trees this one made the strongest and sharpest impression on my mind as well as flesh, pricking its image in me, so to speak' (p. 41). Similarly his passages of natural description are intended to prick the presence of animate nature into the consciousness of the reader. Birds gave him most pleasure as a child, 'not only because birds exceed in beauty, but also on account of the intensity of the life they exhibit' (p. 177). In describing the bird-song of linnets he finds harmony where by rights there should be dissonance, but harmony of a higher order than any produced by mere man: 'It is as if hundreds of fairy minstrels were all playing on stringed and wind instruments of various forms, every one intent on his own performance without regard to the others' (p. 50).

Just as non-human life forms are seen in human terms, so humans are shown to display the attributes of their fellow creatures, especially the birds. This technique is employed with particular intensity in the chapter describing his first visit to Buenos Aires. The large collection of fashionably dressed people that gathers outside the church on a saint's day reminds him of 'a flock of military starlings, a black- or dark-plumaged bird with a scarlet breast' (p. 82). The negro washerwomen covering the beach of the River Plata with their white flapping linen while they chatter and laugh with one another make him think of 'the hubbub made by a great concourse of gulls, ibises, godwits, geese and other noisy waterfowl on some marshy lake' (p. 83). Birds of prey naturally occur to this hater of cities. The raucous cries of the nightwatchmen sound like 'the caw of the carrion-crow' (p. 85). Most of the beggars are ex-soldiers 'thrown out to live like carrion-hawks on what they could pick up' (p. 86). This capacity to discern the animal in the human is exercised, if less concentratedly, throughout the book. Yet it rarely obtrudes.

It is instructive to compare Hudson's book with Jocelyn Brooke's semi-autobiographical *The Orchid Trilogy*, itself a fine piece of writing. In the latter one is immediately made aware of this other naturalist's conscious employment of the orchids which he spent a lifetime collecting as metaphors for virtually every other aspect of his life history. To give just one example, Brooke's first arrival at boarding school, having left behind him his beloved Kent countryside where he had spent his earlier boyhood in search of orchids, is humorously described so as to foreground the presence of orchids there too:

I realized for the first time that there might be compensations for my exile: I should at least find some new flowers. And when we drove up the laurelled drive to the slate-roofed, red-brick building, there, sure enough, was that nice Mr Learoyd to welcome us, with the Green-winged Orchid in his button-hole.[14]

By comparison one hardly notices that Hudson's description of Don Gregorio, for instance, within one page incorporates comparisons of his head to that of a curly retriever, his expression to that of a toad, the sound of his voice to angry barks, and his laugh to the scream of a fox (p. 136). As Nicholas Shakespeare observes, 'Hudson thought of himself as a field naturalist "who takes all life for his field".'[15]

All this is a subtle stylistic preparation for the introduction of animism two-thirds of the way through the autobiography. Throughout Hudson describes his own boyish self as he does all the other humans by similar comparisons to the bird life that captivates his young imagination, a process which culminates in his desire for the wings of a great-crested screamer: 'If I could only get off the ground like that heavy bird and rise as high, then the blue air would make me as buoyant and let me float all day without pain and effort like the bird!' (p. 165). If this longing of his is only occasionally gratified indirectly in dreams in which he is levitating, he does finally forge out of his longing to belong to life in the wild a philosophy of life, a personalized religious belief in the existence of animal and vegetable intelligence. He distinguishes what he means by the term 'animism' from cultural anthropologists who understand it to mean 'a doctrine of souls that survive the bodies and objects they inhabit'. As Hudson first explained this concept in *Idle Days*, for him animism is 'the mind's projection of itself into nature, its attribution of its own sentient life and intelligence in all things'.[16] Hudson first became conscious of 'this sense of the supernatural in natural things' at the age of eight and claims that 'the feeling has never been wholly outlived' (p. 196). His unusual retention of this childlike feeling gave him the confidence to apply his adult perceptions and his immense skill as a prose stylist to the formative experiences of his childhood in his autobiography.

The effect which animism has on him, he writes, is both pleasurable and at times frightening. For the spirit of nature which he had been observing and communing with from infancy struck him as cruel as well as beautiful. The early chapters of the book are scattered with instances of nature's cruelty and violence. There are the 'paroquets' who strip off the peach blossom to make themselves a perch ('it was a desecration, a crime even in a bird' (p. 48)). There are the 'thistle years' when giant thistles cover the plains, restricting the movement of man and beast alike and threatening them with death by fire. Or there is the storm in which half-inch-thick hailstones batter down the crops and kill or injure hundreds of cattle. Deer die locked solidly together by their horns. And the cruelty extends to men. Hudson is both horrified and fascinated by the barbaric behaviour of the gauchos whose heartless method of slaughtering cattle he describes in considerable detail. Nor does he exclude himself from this less pleasant side to nature. Shooting birds for the table he seems to think is justified. Yet when he is pursuing a golden plover, because when cooked it is one of his father's favourites, he recalls his humiliation at being reproved for 'chasing away God's little birds' by, of all people, an uncouth gaucho (p. 151). Just before his discovery of his belief in animism he learns to stop persecuting serpents and is rewarded by a near-mystic vision of a black snake, so representative in his dual nature of all life in the wild. The snake, which had crawled over his foot while he had remained frozen with horror, had left in him 'a sense of mysterious being, dangerous on occasion as when attacked or insulted, and able in some cases to inflict death with a sudden blow, but harmless and even friendly or beneficient towards those who regarded it with kindly and reverent feelings in place of hatred' (p. 190). Hudson continued throughout his early manhood to have ambiguous feelings about killing the birds he collected for a living and sent off to various museums. But his close observation of the workings of nature convinced him that life in the raw necessarily entailed an eternal conflict between predator and prey, nature and man: 'It is a principle of Nature that only by means of strife can strength be maintained. No sooner is any species placed above it, or overprotected, than degeneration begins.'[17]

Hudson sees modern urban civilization as the product of such degeneration, an atrophied state which is infinitely inferior to the risk of death in the wild which makes life there so much more vivid.

The autobiography employs death as a recurrent motif for just this purpose, to help Hudson convey to the reader the intensity of his early life which he shared with the wild life of the pampas. Chapter I associates the house in which he was born with the ghost of a former negro slave who had been scourged to death for daring to declare his love to the mistress of the house. He never saw that ghost, but he did come across an apprehended murderer tied to a post in the great barn of his new home, 'The Acacias', described in Chapter 2. In the next chapter comes the death of his old dog, Caesar, and his tutor's speech over his grave in which he pointed out that we all end up dead and buried like Caesar. The young Hudson is devastated by the thought and immediately sees his condition as similar to that of the murderer awaiting a possible sentence of death. When his mother explains the Christian belief in the immortality of the soul he 'wanted to run and jump for joy and cleave the air like a bird' (p. 33). He is restored to his primitive enjoyment of an unthinking life of the senses. Yet, for all the reassurances that his mother gives him, the frequency of death in his world soon brings back his life-long fear of it, as the rest of the chapter shows. Death threatens the continuation of his wild animal self for which no talk of the soul can compensate him.

The theme recurs in Chapter 8 where the young Hudson is taken by a shepherd boy to see a bloodstain on the ground belonging to an officer of Rosas's defeated troops who had been murdered by his own men. This incident introduces his horrified description of the blood lust of the gauchos who cannot resist slitting a long white neck when they see one, a perversion of nature as Hudson understands it. These repeated references to man's mortality are skilfully drawn together in the coda consisting of the last three chapters. On his fifteenth birthday Hudson is for the first time brought face to face with the end of his boyhood and the likelihood that manhood would mean abandoning his rapturous life amidst nature in favour of 'that dull low kind of satisfaction which men have in the set task' (p. 255). The thought is almost as bad as the prospect of dying itself since both manhood and death mean a loss of his intense immersion in the natural world. Next he falls ill of rheumatic fever and his case is pronounced hopeless by the doctors. He is traumatized at the thought of losing forever his intense life of the senses and feels once àgain like the captured murderer or 'like any wretched captive, tied hand and foot and left to lie there until it suited his captor to come back and cut his throat or thrust him through with a spear, or cut him into strips with a sword . . .' (p. 264), a reminder of the gaucho's method of killing an opponent. He reverts to the terror he felt after the death of his dog Caesar.

But this time Christianity fails to offer him even temporary consolation and he describes his gradual loss of faith following the arguments of a disillusioned old Roman Catholic landowner, his reading Darwin's *The Origin of Species* and his mother's unexpected death. So the book is framed by this motif of death which occurs in the first and last chapters. His abandonment of Christianity has been prepared for in the chapter, 'A Boy's Animism', where he claims that the animistic instinct is usually repressed by Christians with their belief in an implacable anthropomorphic deity continuously watching and judging them, condemning any feelings of the supernatural amidst the natural as sinful temptations of the devil. In *Idle Days* Hudson uses a powerful metaphor to justify his belief in the superiority of the animistic spirit to Christian denials of the barbaric element still lurking within us. This primitive element Hudson compares to 'a hidden fiery core' the heat from which 'still permeates the crust to keep us warm'. This state, he adds sarcastically, 'is, no doubt, a matter of annoyance' to those who wish to be rid of their brute instincts and 'to live on a cool crust and rapidly grow angelic'. Hudson prefers us to be what we are, 'a little lower than the angels'.[18] The combination of geological and Christian imagery here shows Hudson restoring to the natural world the religious feelings appropriated by Christianity.

Eventually Hudson emerges from his own dark night of the soul convinced that his animistic perception of nature will survive his loss of boyhood and that it matters more to him than a belief in immortality. Death is his true enemy because it will bring what he calls his 'earth life' to a complete end. He concludes the book with the bare assertion that throughout his life he has felt 'that it was infinitely better to be than not to be' (p. 289). His spiritual quest has ended with a defence of life on earth that is lived so intensely that material existence reveals the supernatural within itself. In showing that the author has retained contact with his powerful experiences of nature as a child his autobiography acts as a vindication of his subsequent life – especially of his achievement as a naturalist and writer. The childhood vision that led to his conversion to animism in his later teens is responsible, he considers, for the books he has subsequently written and the life he has subsequently lived with the vision undiminished in his mind. So that *Far Away and Long Ago* turns out to be an apology for (which is nearly always a vindication of) a life and belongs to that long line of religious apologies starting with Augustine and stretching to Newman. As with Reid, Hudson's true reader is himself looking for validation to himself as writer for his continuing dependence on the formative drama of a childhood in his case lived in the depths of nature from which his writing draws its inspiration and vision.

John Cowper Powys: *Autobiography*

Like *Far Away and Long Ago*, John Cowper Powys's *Autobiography* (1934) is an apology for a highly unusual life seen as a spiritual quest for imaginative omnipotence over his circumstances. He described it to his sister Marian as 'a sort of Faustian Pilgrimage of the Soul'.[19] He set out to make it a unique contribution to the genre, 'not an *ordinary* autobiography but a very queer one *such as has never been writ before!*' (p. xix). No one would dispute that his autobiography is a completely new departure for which it would be hard to find any earlier models. Many of Powys's admirers rate it his best book. In sheer length alone it is exceptional and depends for part of its effect on its verbal extravagance. The book is packed not with facts about the external circumstances of his life but with the developments of his inner life and his later reflections on them. For Powys the inner life of the imagination is the only true reality: 'Real reality is entirely of the mind. It is partly good and partly evil. But it holds the *lower reality* – that more plastic agglomeration of matter and material force – in subjection. It creates and it destroys this *lower reality* at its wilful and arbitrary pleasure' (p. 28). He sees the 'objective world' as 'really a most fluid, flexible, malleable thing' which is 'rebuilt by each of us in the mind in different form'. (p. 62). He, likewise, is the product of his own 'half-conscious self-creation' (p. 6), and *Autobiography* attempts to trace back to their imaginative origins the components of his later personality.

The autobiography spans Powys's life from his birth in 1872 at

Shirley in Derbyshire to his sixtieth year. The first part of *Autobiography* gives a reasonably straight chronological account of his childhood at Shirley and subsequently at Dorchester and then Montacute in Somerset where his father remained vicar for thirty- two years. Two chapters are devoted to his schooldays at Sherborne, one to his three years at Corpus Christi College, Cambridge, and one to Southwick, Sussex, where he lectured to girls' schools in Hove nearby. In 1896 he married Margaret Lyon, a fact which is barely alluded to in the book. The marriage was not a success and the pair drifted apart after the birth of an only son in 1902. Powys spent much of his life supporting his wife and son from a distance. At this point in the autobiography strict chronology gives way to a more thematic approach to his life between the ages of thirty and sixty. For most of these years Powys earned his living as an itinerant lecturer, dividing his time between Europe and the United States. Hard up for most of the time and suffering from a succession of stomach ulcers, he began producing novels and essays in 1915 and abandoned lecturing for full-time writing in 1929. In 1916 he published with his brother Llewelyn *Confessions of Two Brothers*, a first attempt at autobiography which he intended should be 'one of the most cold-blooded dissections on record, of a living person by his own hand'.[20] In fact it resolves itself into a series of essays on his personal responses to philosophy, morality, literature, friends and the like. His masochism does lead him to offer some devastatingly critical insights into his own failings, but he is too afraid of offending public opinion to introduce, for example, his ruling vice, sexual sadism (in the mind), a major motif in his later autobiography.

Autobiography comes at the end of the period of five years during which Powys retired to 'Phudd Bottom' in upper New York State to write full-time. The book was written at speed between August 1933 and May 1934. Like Hudson he relied almost exclusively on memory and let chronology and dates go to the devil, as he put it to his brother Littleton.[21] Although he spent as much time revising as writing each chapter, the shape of the book reflects its spontaneous composition. The result is a kind of deliberate shapelessness which he cultivated in an attempt to capture the evanescent quality of life experienced from moment to moment:

It is most important in writing the tale of one's days not to try to give them the unity they possess for one in later life. A human story, to bear any resemblance to the truth, must advance and retreat erratically, must flicker and flutter here and there, must debouch at a thousand tangents (p. 237).

This is Powys's way of seeking to avoid what Georges Gusdorf calls the original sin of autobiography, retrospective rationalization. Even the conclusion is a beginning rather than an ending: 'It seems as if it had taken me half a century merely to learn with what weapons, and with what surrender of weapons, *I am to begin to live my life*' (p. 652). Belinda Humfrey goes so far as to call *Autobiography* an anti-autobiography which parodies the traditional use to which the genre has been put.[22] Although Powys is bound to flout a number of autobiographical conventions in his determination to write something totally unprecedented, to call it a parody is to attribute more artistry to the book than his approach allowed. He was more concerned to make his book completely *different* from any other autobiography. He also aimed to make it a distinguished contribution in its own right – 'if this Book lasts,' he wrote to Littleton, '& I am trying to make it good enough for that!'[23]

One of his most unusual decisions was to omit all mention of the major women in his life, although he is not entirely consistent in this resolve, and felt free to include the sylph-like girls on whom he projected his sadistic fantasies. He claims in a letter to Littleton that his motive for this omission is to avoid hurting anyone's feelings, especially those of his separated wife. Timothy Hyman may also have a point when he suggests that physical sexuality was of small importance to Powys compared to his fantasies about sylphs.[24] At the same time what he calls 'this *Grand Cut*'[25] leaves him free to explore the feminine in himself: 'In my non-human cult for impossibly slender sylphs I resolve myself – like all true contemplative ecstatics – *into the element I contemplate*' (p. 275). His attitude to the female sex in general is extraordinarily complex and at one time sweeps him up in a wave of revulsion from any manifestation of it, even in his bitch, a phase he depicts with savage irony at his own expense.

In return his rule of exclusion offers him the opportunity to pay far more attention to his own follies and vices. He is determined to save everyone's feelings except his own. His friend Louis Wilkinson wrote a brilliant and convincing analysis of the masochism underlying the sadistic fantasies he parades in *Autobiography* in *Welsh Ambassadors* (1936). Wilkinson argues that Powys's real pleasure in, for instance, reading sadistic books 'came from his foreknowledge of the self-punishment that he was going to inflict by forbidding himself to read such books at all'.[26] Wilkinson cites such passages in *Autobiography* as that in which Powys describes how he admitted to an American audience to whom he was lecturing during the Great War that he was not fighting because of his fear of German bayonets (p. 601). What Powys omits to tell his audience is that he had twice tried to enlist with the armed forces and on both occasions had been turned down on medical grounds. Self-punishment holds an irresistible attraction for him. In the book he finds excuses to indulge it in describing his physical comforts, his standing in his friends' eyes and his sexual fantasies alike. In leaving himself sufficient space by his Grand Cut in which to pillory himself at each stage of his life, Powys was in fact contriving to satisfy yet again his masochistic needs in the very act of writing his autobiography.

At the same time there is something of the saint about him and in the autobiography he explains his masochistic behaviour as the product of 'an embryo-saintly desire, to return blessing for cursing' (p. 338). By arbitrarily eliminating so much of his external life Powys is able to pay far more attention to his internal contradictions than is normally possible to the conventional autobiographer. He insists that he is a Pluralist, not a Monist (p. 55). He shows a real determination to avoid simplifying his past self and lists five discordant elements all fighting with each other in his early compound personality – a desire to enjoy the Cosmos, to appease his Conscience, to play the Magician, to play the Helper and to satisfy his Viciousness. The list illustrates the way in which he views his life as a spiritual struggle, 'the struggle of a soul, conscious or half conscious, with the obstacles that hinder its living growth' (p. 46). He compares the soul to a native spring that has to thrust itself clear of accumulated debris and the constant invasion of a tidal estuary – both forms of external reality. His explanation of the simile is clearly reminiscent of Bunyan: 'From the beginning madness and fear beset us, vice enthrals us, humiliation benumbs us, pride intoxicates us' (p. 39). But he does not share Bunyan's faith in a Christian God. He refers only to a First Cause that he blames for all the suffering in the world.

Yet he is convinced that he has what he calls a 'soul'. For his soul to develop or find itself it has to have a goal or ideal pattern, what Powys calls his 'life-illusion'. He borrowed this expression from Ibsen. In *A Philosophy of Solitude* (1933) he defined it as 'that secret dramatic way of regarding himself which makes [someone] feel to himself a remarkable, singular, unusual, exciting individual . . . It is the shadow of your subjective self.'[27] One's life-illusion is formed by the creative power of the imagination which enables an individual 'to substitute a secret reality of his [own] for the reality created by humanity' (p. 85). The private blueprint therefore elevates the inner life of fantasy above social life in which Powys was such a misfit. One's life-illusion is far more real for him than the illusion of 'space-time'. He sees it as 'within the power of the will and the imagination to destroy and re-create the world' (p. 26). He thinks that his power is what earlier generations had externalized as 'God':

Never mind the name! The point is that we *have* the power of re-creating the universe from the depths of ourselves. In doing so we share the creative force that started the whole process. It is personality, the outrushing energy of living organisms, that underlies all the criss-cross currents of the world (p. 361).

Powys's philosophy of life, then, boils down to a cult of personality seen as 'the living well-spring of mysterious magic in us' (p. 361). Imagination is at the source of individual personality as it is the source of the magician's powers. This is why the figure of the magician plays such a central role in *Autobiography*. Powys wants to reintroduce the child's belief in the efficacy of magic into adult life. On the opening page he instances his imaginative power as a child in converting Mount Cloud, a local hill, into an immense mountain as a form of natural sorcery which as adults we spend a lifetime trying to regain. He also remembers as a toddler announcing to his nursemaid that he was 'the Lord of Hosts'. He genuinely believed that he possessed all the Lord's powers over the rest of creation. He gives a number of instances of his childhood fetish worship which he carried with him into later life. In boyhood his longing to possess supernatural powers took the more conscious form of inventing a mythology of his own with imaginary races and an invented language. Although school didn't encourage his cultivation of magical powers he continued to believe in his ability to tap 'a deep reservoir of magnetism' which gave him power over animate and inanimate life forms (p. 225). In later life he even felt constrained to curb his feelings of malignance towards his enemies because of the number of occasions on which his enemies met with overwhelming disaster (p. 352). He also held his magical powers responsible for his success as a lecturer in which role he became the author he was lecturing on, 'like a demon possessing a person' (p. 457). His cult of childhood, then, is another version of the Romantic's longing to repossess the spiritual purity and wholeness of one's youth.

Repeatedly Powys compares himself to Taliesin, the Welsh magician-cum-bard of the sixth century who knew all the secrets of the past, present and future. Taliesin claimed to have witnessed the whole of human history from the fall of Lucifer onwards and prophesied that he would endure to the end. Some of his poems in the Book of Taliesin describe some of the things he had been in the past including both inanimate objects and animals. Powys, who claimed descent through his father's family from the ancient Welsh princes of Powysdom, felt strongly drawn to Wales well before he finally settled there to live out his old age. He consciously modelled himself on Taliesin from his childhood onwards. In his early thirties, he records in *Autobiography*, finding himself to be 'instead of a conscientious scholar, an imaginative charlatan', he resolved to realize with all his spiritual force his inheritance from the ancient Druids. In the same passage he anticipates his later role 'as a restorer of the hidden planetary secrets of these mystical introverts of the world' (i.e. the Druids) (p. 335), a reincarnation of Taliesin. Taliesin's thaumaturgic powers of becoming other people and things are what Powys most aspired to in his own life. What came naturally to the young child had to be relearnt by the adult lecturer and writer. Possessed by the spirit of Taliesin, Powys claims: 'I could become inanimate objects. I could feel myself into the lonely identity of a pier-post, of a tree-stump, of a monolith in a stone-circle; and when I did this I *looked* like this post, this stump, this stone' (p. 528). There is therefore a direct connection between Powys's cultivation of magical powers and his approach to literature, especially to his own autobiography. Late on in the book he makes this general connection: 'What we need – and the key to it lies in ourselves – is a bold return to the *magical* view of life. I don't mean to the magic of Madame Blavatsky, but to that kind of faith in the potentialities of the ego, with which all great poetry and all great philosophy has been concerned' (p. 626).

Jeremy Hooker sees the development of Powys's magical view of life as the main theme of the autobiography.[28] Powys certainly draws on his thaumaturgic powers to become not just the other characters in *Autobiography* but also his past selves. At the same time Powys in his role as magician constantly comes close to madness of a kind. He is the first to admit how nearly mad he has been for most of his adult life. Nevertheless he has quite deliberately cultivated his seeming insanity as a way of protecting his life-illusion. He first discovered the advantage of pretending to be mad at Sherborne when challenged to throw a sponge at a notorious bully's head and used this ploy also to escape the trapeze and horizontal bars in the gymnasium. Later on in life he employed the same device to nonplus those who insisted on the supremacy of scientific truth or of academic knowledge: 'When I am in their company I cannot describe to you the voluptuous pleasure I derive from struggling to be a greater fool than Nature has made me' (p. 428). Here he is simultaneously capitalizing on his masochistic streak. He claims to get 'a thrill that *is* actually sensual' by making a fool of himself in this way (p. 427). He calls the malice which causes his displays of pseudo-madness 'sacred malice', or 'Cowperism' after his half-mad poet of an ancestor. His 'Cowperism' or 'mania for acting the zany . . . contains', he asserts in the final chapter, 'a quite definite philosophy of life . . . that combines reaction with revolution in a way in harmony with Nature's own devious and yet magical method of going to work' (p. 641).

'Cowperism' lies at the heart of Powys's autobiographical strategy. It enables him to indulge his fetishes of objects without excuse. It satisfies his masochistic need to make a fool of himself before the world. It offers him the means to escape the boring requirements placed on conventional autobiographers and to portray instead areas of the soul that most of us never admit to even to ourselves. It is his 'organ of research' in his investigation into his spiritual history. 'And what', he asks, 'is the use of these clever sophisticated "Memoirs" of tedious dignitaries, who never once let you see the shivering, jerking, scratching, crying, groaning, God-alone-sees-me nerve of their central "Libido"?' (p. 603). His book is a celebration of the seeming madness of the magical in the midst of everyday life.

Autobiography is in a class of its own partly because Powys does actually derive satisfaction – both psychological and philosophical – from parading his neuroses, phobias and darkest fantasies. His 'sacred malice' is tapped within the book to further its truthfulness. His most frequent way of turning it on himself is to use caricature paradoxically to heighten the authenticity of his self-portrait: '*Caricaturing* is the master-trick! And that is why the discreet, dignified, plausible autobiographies are so insipid and unconvincing. A touch of caricature is what we *must* have, if we are to compete, even in this analytic job, with the beautiful madness of Nature' (p. 642).

Near the end he quotes Blake's aphorism, 'Excess is the path to Wisdom'. Powys uses excess in almost every aspect of *Autobiography* – its extraordinary length, its circumlocutory style, its satirical self-portrait, its outline of the protagonist's zany beliefs. He prefers to sacrifice what passes for autobiographical 'truth' so as to be free to pursue what he calls 'wisdom'. To find that he has to combine the powers of the magician, the imagination of the poet, and the sacred malice of the near-madman. His autobiography flies in the face of conventional expectations of what its contents and organization should consist. It dares more than most. It can be seen as the idiosyncratic outpourings of an egocentric and demented eccentric, or – and this is how I see it – as one of the most startlingly original portraits of the inner or spiritual self to have emerged this century.

10

The Double Perspective:
The Self and History

All the autobiographies that have been given extended consideration
in this book have been selected because they give priority to a
subjective view of the self. This is only partly a matter of perspective.
If the autobiographer shifts his main focus from himself to actions he
has witnessed or to people he has met then he is moving into the
sub-genres of memoir and reminiscence, or associated genres like
travel writing or the philosophical essay. Such writers assume that
the scenes or people they have met with will be of more interest to
their readers than themselves. So they seek to bend the genre to their
purposes by blending it with other forms of writing less focused on
the personality and identity of the writer. But one of the distinctive
features of autobiography is its natural tendency to reveal more of
the character of the author than he is aware of – by the material he
has selected as significant, by the way in which he has ordered it and
given it shape and therefore significance, by the style and language
which he employs for his self-presentation, and by the manner in
which he portrays himself and others. What I have called the
subjective autobiographer, on the other hand, knowingly embraces
these potentialities within the genre, committing himself to a higher
level of exposure than he can consciously keep within his control.
Where the danger for a writer of a memoir or reminiscence is that of
disappearing behind the people, places and events he has witnessed,
the danger for the subjective autobiographer is that of everything else
disappearing behind his concentration on his past self. But, as Erik
Erikson has shown, autobiography concerns itself with three forms
of time – the writer's present, the writer's past, and the historical
context in which the writer has lived.[1] It is this sense of historical
context that distinguishes the autobiographer who accepts the
phenomenological fact that he is the centre of his world, the focus

point of any experience, from one whose narcissism makes him believe that he is also the only worthwhile object of all experience and therefore tends to represent himself as untouched by social or historical circumstances.

In *Design and Truth in Autobiography* Roy Pascal, who also gives priority to subjective autobiography, insists that it should be 'the story of a life in the world' which offers us 'events which are symbolic of the personality as an entity unfolding not solely according to its own laws, but also in response to the world it lives in'.[2] This double perspective, as I term it, what Pascal calls 'the balance between the self and the world, the subjective and the objective',[3] characterizes for him the best examples of autobiography. Certainly it avoids the discursiveness of the reminiscence and memoir on the one hand, and the tendency to narcissism on the other. Gosse's *Father and Son* is an instance in point where the struggle he describes is not simply one between two temperaments, but, as he announces at the start of the book, also one between two epochs. Here is an autobiography that advertises itself as being as much about the father as the writer-son and as much a 'diagnosis of a dying Puritanism'[4] as a personal history, that nevertheless manages to make of his subjective self the focus and controlling medium of all these factors outside himself. Yeats, Graves, Sassoon, Lee – so many of the most satisfying autobiographers employ this double perspective that one might be tempted to accept Pascal's prescription and look for both ingredients in all worthwhile subjective autobiographies. But in the first place Pascal is convinced that in general modern autobiography suffers from 'a general lack of relationship between personal and social being'.[5] Yet this is not true of most of the autobiographies considered so far in this book. In the second place how is one to account for the excellence of such modern autobiographies as Conrad's or Reid's in which historical and social influences play a relatively minor part? Both nevertheless manage to counteract the temptation of narcissism by a use of formal literary devices which help to convince the reader that the writer is firmly in control of his material and not indulging an egotistical bent without regard to the reader's needs and satisfactions.

On the other hand there are some autobiographers who set out to portray the interaction of the self with its society and times and for whom this is central to their conception of autobiography. Such autobiographers tend to see their identity as their very particular mode of relating to the world around them. Typical of this approach is Vera Brittain's acclaimed autobiography of her earlier life during the first quarter of this century, *Testament of Youth*, in which her stated intention is 'to put the life of an ordinary individual into its niche in contemporary history, and thus illustrate the influence of world-wide events and movements upon the personal destinies of men and women'.[6] The influence can be two-way and doesn't have to make of the individual a victim of historical determinism. At least three of the four autobiographies on which this chapter concentrates show the protagonist resisting forces of history and discovering his own identity in the course of doing battle with society. But for all of them the central problem is to what extent they are a product of local and temporal conditioning, or, alternatively, how autonomous is the inner self which detaches itself from outer events and which also constitutes the reflecting self of the autobiographer. The four autobiographies offer an interesting range between the most determinist and the most interior response to the twin forces contributing to the autobiographer's sense of identity.

H. G. Wells's *Experiment in Autobiography* offers an example of this sub-genre taken to the extreme of exteriority. He tries to see himself solely as a representative brain born at a particular moment in the development of civilization. What saves him from himself is his inability consistently to maintain this impersonal stance. Arthur Koestler's two volumes of autobiography, *Arrow in the Blue* and *The Invisible Writing* are like Wells's book in seeking to portray their protagonist as a representative mid-European man in the earlier part of the twentieth century caught up by some of the major historical movements of that period. In Koestler's case his keen psychological insights into his make-up and his insistence on a supra-conceptual level of reality are consciously used by him to maintain a balance between the self and the self's apprehension of the impact on it of the world in which it exists. Stephen Spender's *World Within World* shows a bias towards the personal end of the spectrum and only allows the world to intrude into his concentration on the self during his account of his life in the thirties when his private life was invaded by political events. Finally Storm Jameson's *Journey from the North* stands at the opposite extreme to Wells's autobiography. She insists on personalizing history, depicting the historical periods through which she has lived in the restricted focus of her individual, searingly felt experience. Despite their considerable differences in emphasis all four autobiographies show their authors searching for a way of reconciling an interior view of the self with a recognition of the way in which the self has been shaped by the world outside it. In all of them the double perspective is a predominant feature which largely determines the material selected and its manner of presentation.

H. G. Wells: *Experiment in Autobiography*

H. G. Wells began writing *Experiment in Autobiography* in the spring of 1932 at the age of sixty-four. Wells was in a trough of his life at this time. His wife and business manager, 'Jane', had died four and a half years earlier. His long-term affair with Odette Keun was acrimoniously breaking up. He was convinced that he was being treated increasingly as a writer of a past age, an anachronism in the era of Auden and Orwell. In the Prelude to his autobiography he talks of this 'phase of fatigue and . . . discouragement'[7] and gives as a major motive for undertaking the book the hope that by clarifying these discontents he will rid himself of them or at least bring them within his control. By the final pages, which he wrote in mid-1934, he was able to assert that he had achieved his purpose of reassuring himself, mainly by reaffirming his belief in a world state: 'The stream of life out of which we rise and to which we return has been restored to dominance in my consciousness, and though the part I play is, I believe, essential, it is significant only through the whole' (p. 825). His sense of identity, then, is heavily dependent on the validation he receives from society. He actually compares his submission to a greater order to the reaffirmation of an earlier religious mystic; only his modern belief, he claims, does not involve the mystic's escape from fact. His 800-page-long autobiography spans the full range of his sixty-eight years from 1866 to 1934, and incorporates events that were happening in his life during the two and a bit years he spent writing the book. When Gollancz published it in the autumn of 1934 in two volumes Wells was furious because he had intended it as a unity.[8] He subsequently referred to it as 'a very frank and explicit autobiography'[9] and wrote a supplement called 'Exasperations: The Last Testament of H. G. Wells' in 1944–5 (he died in 1946) which was published posthumously only in 1983. It shares the bitterness of his last published book, *Mind at the End of its Tether*, and could not have been incorporated in *Experiment in Autobiography* without ruining its existing shape and purpose. His published autobiography, then, is a construction that reflects a moment of time in his life when, partly by writing the book, he reacquired for a while a belief in himself as an agent of the worldwide progress towards a higher state of civilization.

As if unaware of the extent to which his current self was stamping its presence on the book, Wells as narrator chose to adopt the tone of an intellectual researcher in an attempt to distance himself as author from himself as protagonist. He loftily talks of his autobiography as 'the history of the adventures of a sample human brain in the latter phase of the Private Capitalist System' (p. 159). More specifically the book traces the history of his brain from its narrow origins to its ever-widening horizons ending in the final chapter entitled 'The Idea of a Planned World' with his belief in and advocacy for a world state. One of the many enigmas inherent in the book is Wells's confident assertion in the same opening chapter in which he reveals his current discontents and need to master them that the autobiography will 'culminate in the achievement of a clear sense of purpose, conviction that the coming great world of order is real and sure' (p. 28). Is he genuinely suffering from doubts concerning the certainty of man's scientific progress, doubts that finally overwhelmed him during the Second World War? If so how can he be so confident that by the time he has reached the end of the book he will have overcome these doubts? Or has he inserted his sense of assurance retrospectively after getting to the end of the narrative? He claims that *Experiment in Autobiography* 'is not an apology for a life but a research into its nature' (p. 419). Yet he appears to vacillate between these two positions. The tension created by the conflicting demands of each perspective is responsible for a great deal of the book's interest.

The 'Experiment' of the title refers partly to the attitude he adopts to himself as subject of the autobiography. He wants to see himself as a laboratory specimen: 'I am my own rabbit because I find no other specimen so convenient for dissection' (p. 417). This means that he is constantly trying to see his own particular life as representative of the changes taking place in his civilization during his lifetime: 'My story therefore will be at once a very personal one and it will be a history of my sort and my time. An autobiography is the story of the contacts of a mind and a world' (p. 28). This makes him picture even his parents as 'economic innocents made by and for a social order . . . that was falling to pieces all about them' (p. 59). If he had written entirely in this generalizing vein the book would have been tedious. In practice Wells draws on his wider vision of the significance of personal events to give shape and purpose to a particularized account of his life. He vividly evokes, for instance, the poverty and squalor of 'Atlas House', the china and crockery shop in Bromley High Street that was his home for the first thirteen years of his life. He might have progressed 'from Atlas House to the burthen of Atlas' (p. 417), as he put it, but on the way he offers the reader numerous fascinating details concerning the drudgery of shop and home life. All of the changes to his life, which have quite particular origins, he saw as typical of his time: 'I was following a road along which at variable paces a large section of the intelligentsia of my generation was moving in England, towards religious scepticism, socialism and sexual rationalism' (p. 186).

In relating the particular to the general in this way, Wells is aware of the danger of appearing egocentric and pompous. 'It is unavoidable', he explains in the final chapter, 'that at times I should write as if I imagined that – like that figure of Atlas which stood in my father's shop window – I sustained the whole world upon my shoulders' (p. 643). That, he feels, is the condition of his generation. But more than once he parodies this autobiographical approach. For instance the top hat he wore on Sundays to take out his cousin and future wife, Isabel, gives rise to a humorous disquisition on the rise and fall of his successive top hats. 'They mark periods in human history', he declares with mock solemnity, 'as surely as do the ramshackle houses in which I spent the first half of my life and the incoherent phases of my upbringing and education' (p. 284). Passages like this one indirectly acknowledge his recognition that his continual reference of his personal predicament to the wider context of historical change tends to turn him into a depersonalized socially-determined representative of a passing phase of civilization without control over his own destiny. He resists this interpretation not merely by indirect self-mockery but by direct confrontation with his private and often selfish reasons for conforming (or not) to one of these norms. Speaking of his advocacy of free love when he was in his thirties, he writes that a 'few tactful omissions would smooth out the record beautifully' (p. 435). But he insists on the truth: 'I could preach these doctrines with no thought of how I would react if presently my wife were to carry them into effect, since she was so plainly not disposed to carry them into effect' (p. 436). He also reminds himself of the storm of irrational jealousy and possessiveness that overtook him on a return visit to his estranged first wife, Isabel. Each representative phase of his life is similarly re-examined in this way from the private underside. So that his autobiography plots the struggle that took place in him between his rational belief in advancing the state of civilization and his emotional resistance to an advance which he persuades himself is historically inevitable.

There are also moments in the autobiography when Wells stops in mid-track, acknowledges that he has been oversimplifying the picture and starts anew as if, he says, he 'were a portrait painter taking a fresh canvas and beginning over again' (p. 419). One such moment comes half-way through the book after he has described his growing alienation from his first wife, Isabel, who came to represent a primary set of values (obligations to his immediate world) and his attraction to Catherine Robbins with whom he associated a secondary set of values (opposed to his immediate obligations, offering a freer, wider life). He even asserts that 'the primary theme of this autobiography is this conflict between the primary and the secondary values of life' (p. 364), between the old eighteenth-century order and the new vision of a world state. Then at the beginning of the next chapter (illuminatingly called 'Dissection') he repudiates this simplified version of his relations to the two women. In order to write the autobiography Wells recovered from his friends some thousand of his letters. The memories these revived in him brought him face to face with his complete sexual confusion at this time of his life. He admits that he had even persuaded his biographer, Geoffrey West, to represent his divorce as part of his progressive detachment from the narrow circumstances of his origins. Now the letters force him to realize that if separation happened to help him detach himself from primary values at this moment, later his sexual drive (for instance, his continual hankering after Isabel) was to hamper his progress towards a new vision of life based on secondary values.

Another minor volte-face of this kind occurs when he makes a comparison between his own neglect as a writer and prophet and that of Roger Bacon. No sooner does he write of this comforting fantasy than he realizes how absurd and unreasonable it is to imagine that all the millions of words he has had printed have had no effect on anyone. What is interesting about this moment of realization is that it shows the extent to which these checks on his generalizing tendency arise out of the act of writing about himself. Autobiography might produce 'inflammation of the ego' (p. 729), but attention to its self-reflective nature ensures that self-criticism promptly counters its initial temptation to indulge in narcissism.

I said that the title of this autobiography refers partly to his experimental attitude to his past self seen as a laboratory specimen. But *Experiment in Autobiography* is also precisely that, an experiment in the use of the genre. Wells's self-conscious employment of the full potentialities of autobiography ensures that he keeps both perspectives, individual and social, in continuous conflict with one another. Primary and secondary values are repeatedly defeating one another, making the composition of the book reflect the irregularity of his life. 'If you do not want to explore an egoism,' he tells the reader, 'you should not read autobiography' (p. 417). Just as the book is about his private and public selves, so it is written both for himself and with the reader as critical observer in mind. If anything, he remarks, his concentration on life's general patterns, which governed his earlier adult life, has been replaced by an interest in the individuality of personalities in his maturer years. The autobiography attempts, not wholly successfully, to mirror this belief in the series of portraits (of Roosevelt, Balfour, Lenin and Stalin) which he offers near the end of the book.

But his self-portrait is in its complexity and internal contradictions superior in my opinion to any of his fictional characters. Wells readily admits the extent to which, despite his determination to be as true as he can to the inconsistencies of his character and actions, his self-portrait is an oversimplification of the reality. In fact, he writes, he has done no justice to the 'countless subsidiary happenings, to the fun of life and the loveliness of life and to much of the oddity of life' (p. 822). This tendency to make his self-portrait excessively 'bony and bare' is an unavoidable by-product of the genre's pressure to create shape and meaning. He can counter it only by drawing attention to its presence. Nevertheless he regards autobiography as potentially superior to the novel once the writer has been freed of prevailing social and moral inhibitions. For him it is more suitable for 'the analysis of individual difference' if it is allowed to achieve its full potentiality as a 'searching and outspoken' medium (p. 502). Since he analyses his own individual difference in most detail, what he seems to be arguing here is that autobiography is a superior instrument for *subjective* self-analysis. Despite his claims to be no more than an objective researcher plotting his earlier self's pattern of life within the wider pattern of the history of mankind, he ends up championing the genre's superior capacity to depict the inner self, both past and present. Whether Wells is right or not in his assessment and his prediction that autobiography will replace fiction at some future date, his exploitation of the contradictory potentialities within the genre is at the root of the continuing durability of his own experimental autobiography.

Arthur Koestler: *Arrow in the Blue*; *The Invisible Writing*

Towards the end of *Experiment in Autobiography* Wells justifies his gallery of the politically famous whom he had met with an aphoristic statement: 'The more completely life is lived the more political a man becomes' (p. 781). The circumstances of Arthur Koestler's life were such that politics entered it from childhood not by choice but as an unavoidable determining factor. Perhaps his lonely childhood, much of which he spent in hotels because of his mother's neurotic antipathy to running a normal home, made him ultra-susceptible to outside influences. From his birth in Budapest in 1905 to his arrival as a penniless refugee in Britain in 1940 he was swept along by the forces of change that produced two world wars, the Communist Revolution, the destruction of his country's independence and the extermination of most of his mid-European friends and relatives. As a Jew he spent several years in Palestine before and after the creation of the state of Israel. His fluency in German and French took him to Paris and Berlin as a foreign correspondent. His eye-witness experience of Hitler's rise to power in Berlin led him to join the Communist Party from 1932 to 1938 and was responsible for his three-month ordeal in Franco's gaols where he expected to be executed. On the outbreak of the Second World War he was interned by the French and escaped the Nazi occupation only by joining the French Foreign Legion for three months. By the time he reached London in 1940 where he spent the first six weeks in Pentonville gaol, he justifiably felt that his individual life could be seen properly only in the light of the political events which had so affected his own destiny.

It is not surprising therefore to find that in all four of his autobiographical books he emphasizes his role as a representative political animal of his time. Two of these books were written immediately after the events they describe took place. On his release from Seville gaol he dictated to his first wife, Dorothy, *Dialogue with Death*, his account of the six months he spent in Spain in the first half of 1937, first as a correspondent, then as an imprisoned spy. Originally published together with documentary material from a propaganda book he wrote after a previous visit to Spain as *Spanish Testament* (1937), it was reissued on its own in 1954. Even at this date he shows a marked leaning towards a determinist view of human personality: 'how little depends on what a man is, and how much on the function which society has given him to fulfil'.[10] His next autobiographical book, *Scum of the Earth* (1941), was his first book to be written in English. It was hastily put together after his release from Pentonville and covers his life in France from August 1939 to August 1940. In escaping the Nazi holocaust Koestler considered himself atypical and gives this as his reason for ending the book where he does. Once again he sees as the justification for indulging in autobiography the fact that 'his personal adventures were typical for the species of men to which he belongs: the exiled, the persecuted, the hunted men of Europe; the thousands and millions who, for reasons of their race, nationality or beliefs, have become the scum of the earth'.[11]

In his late forties Koestler, 'before saying farewell to politics', as he explained in *Bricks to Babel*, 'felt the urge to take stock of the past'.[12] By the early 1950s he was ready to turn his back on politics in order to devote the rest of his life to writing about science and philosophy. He was simultaneously undergoing a crisis in his own life during this period: his marriage to his second wife, Mamaine, broke up and his application to take up permanent residence in the United States was blocked, causing him to move to England for what turned out to be the rest of his life. The result of his stock-taking took the form of two volumes of autobiography, *Arrow in the Blue* (1952), covering his first twenty-six years from 1905 to 1931, and *The Invisible Writing* (1954), continuing his story from 1932 to 1940 with a brief Epilogue taking him to 1952 when he finally settled in England for good. After completing this diptych he publicly renounced his role as political pundit in the Preface to a volume of essays, *The Trail of the Dinosaur* (1955): 'Cassandra has gone hoarse, and is due for a vocational change.'[13] The second part of his life he spent writing against positivist and materialist interpretations of the natural sciences, and he also showed an increased interest in parapsychology. Towards the end of his life he tried to demonstrate, especially in *Bricks to Babel*, the unity underlying the two phases of his life and work, a unity already demonstrable in embryo in his two great volumes of autobiography. In 1983 Koestler, suffering from Parkinson's disease and leukemia (CCL), committed suicide, as did his third wife, Cynthia. His publisher had persuaded him in his last years that he could best demonstrate the unity in his life through a continuation of his autobiography. He left behind him both a series of short autobiographical memories he called 'Micromemoirs' and an unfinished manuscript of an autobiography written with Cynthia, each undertaking alternating chapters, covering the years 1940 to 1956. The latter was published as *Stranger on the Square* in 1984.[14]

At the end of his two-volume autobiography Koestler refers to himself up to the age of thirty-five as 'this typical case-history of a central European member of the educated middle classes, born in the first years of our century'.[15] More than one commentator has mistakenly taken Koestler at his word and then proceeded, like Roy Pascal, to condemn him for his failure to portray the whole man including the inner self.[16] It is true that both Wells and Koestler show a similar fear of narcissistic self-indulgence in their approach to autobiography. Yet simultaneously both are at least as much concerned to portray their inner as their outer selves. In fact both partly hide from themselves as well as the reader their genuine fascination with their own individuality. Both are also too honest not to admit at certain points in their autobiographies to the partial atypicality of the 'representative' protagonists. On the one hand Koestler is constantly aware of 'the danger of sincerity degenerating into exhibitionism' (II, p. 361). This, he writes, is what happened to his abandoned first attempt to write an autobiography franker than Rousseau's. On the other hand he claims to feel nearer to Rousseau than to any other autobiographer in that Rousseau's ' "whole life" ', he quotes, ' "had been an attempt to be himself and nothing else besides" '[17] – just like his own, he adds.

Like Wells, too, Koestler makes highly self-conscious use of the autobiographical medium. The third chapter of *Arrow in the Blue*, called 'The Pitfalls of Autobiography', is spent discussing the relative claims of concentrating on the internal or the shared public events in an individual's life, of sincerity against art, and goes on to ask what class of reader the autobiographer should address himself to. His answer to the last of these cruxes is a decision to direct his autobiography to 'the unborn, future reader' (I, p. 27), which indicates the ambitious nature of this two-volume autobiography. In *The Invisible Writing* he compares his original externalized autobiographical account of his Spanish experience, *Dialogue with Death*, written when he was thirty-two, with the chapter explaining those same events fifteen years later. At the end of the chapter he confesses that his revised version of his spiritual crisis at this time is itself 'far too tidy and logical' compared to the changing and uncertain actuality (II, p. 357) and speculates on how different again a further autobiographical account of that experience would be if written fifteen years on. He is very alive to the temporal and emotional relativity of the genre, especially where it employs an interior as well as an exterior perspective: 'In the realm of psychology no concrete, objective truth exists, only an almost infinite number of levels of truth' (II, p. 362). Koestler labels the exterior viewpoint 'the Chronicler's urge' and the interior 'the *Ecce Homo* motive'. He advocates a synthesis of the two perspectives but is conscious of the tendency of most autobiographers to fall into the extremism of one polarity or the other: 'The vanity of men in public life detracts from the autobiographical value of their chronicles; the introvert's obsession with himself makes him neglect the historical background against which he moves' (I, p. 24). It is the second danger which made him break off his autobiography at

1940, after which point he felt that his life no longer interacted so directly with the mainstream of political events.

At first sight the reader might be tempted to see Koestler's political life and thoughts as representative of the exterior perspective and his psychological analysis of his motives and impulsive actions as representative of the internal perspective. Take his account of his six years between 1932 and 1938 as a member of the Communist Party as an example. He spends considerable time analysing the circumstances which led to his joining the Party. His conclusion is that his 'progress toward the Communist Party followed a typical, almost conventional pattern of that time' (I, p. 260). He resorts to an analogy from physics to represent himself as one among millions east of the Rhine drawn like iron-filings towards either Fascist or Communist fields of force. Nothing could sound more determinist. Yet he also offers a psychoanalytical explanation for his conversion. In effect he was a neurotic suffering like most rebels and revolutionaries of his generation from a need to escape from the collapsing world of his parents, motivated by frustration, anxiety and guilt. Even his Communist mentors are shown to fulfil an analyst's function by encouraging transference on the part of the convert cum patient. Yet it must be obvious that these psychological explanations are just as generalized as the political ones. Koestler leaves the reader in no doubt that retrospectively he can see no difference between the closed systems of Communism, Freudianism and Catholicism. Quoting from his contribution to *The God That Failed* (1950), he suggests that both the Historian's and the Psychiatrist's views of the revolutionary reflect half-truths; both seek to explain individual action as symptomatic of universal patterns of behaviour.

Does this then imply that for Koestler individuality is illusory? How about, for instance, the childhood origins of his adult conversion to Communism? When he was fourteen he lived through the heady hundred days of the Hungarian Commune of 1919. Just before the Communists came to power he witnessed the mass funeral procession following the coffins of some Communist martyrs which was accompanied by a band playing Chopin's Funeral March. 'Chopin's March', he writes, 'made a romantic Communist of me long before I knew what that word meant' (I, p. 62). Similarly he describes how his initiation into a Zionist duelling fraternity in Vienna at the age of seventeen was responsible for his twenty-five year involvement in the struggle for a Jewish state. A third incident in which temporary panic caused him to flee Madrid in 1936 left him with such a residue of shame, he writes, that it led him to assuage his guilt by foolhardily staying on in Malaga after its capture by Franco's troops was a certainty, which in turn brought him face to face with death. That experience was responsible for his break with the Communist Party. In each of these cases Koestler appears to be attributing purely personal and accidental causes for his becoming part of the great historical forces of his time. But he dispels this impression early on in *Arrow in the Blue* by explaining that political commitment invariably begins with emotional commitment: 'The arguments which justify and rationalize the credo, come afterwards,' he asserts (I, p. 99). To clinch his argument he compares emotional conversion to the primitive infantile emotive phase and rational justification to the ascendancy of the reality principle over the pleasure principle. This use of Freudian terminology is typical of his strategy throughout the two volumes by means of which he uses one closed system to correct the half-truths of another.

Nevertheless Koestler does believe that it is possible to escape all such closed systems, as the titles of both volumes indicate. In retrospect he sees that he and his contemporaries were brought up to be enlightened and reasonable just when 'the age of Reason and Enlightenment was drawing to a close' (I, p. 235). But opposed to the widespread expectation passed on to him at his *Realschule* that all the mysteries of the universe would ultimately be solved by science and reason was his realization in 1919 that an arrow shot into the blue would enter infinity and eternity, concepts which the old science refused to recognize as actualities. An attraction for mystic unreason, which produces in him a heightened state of consciousness in which the ego dissolves into the universe, invariably lies behind some of the most apparently suicidal gestures in his earlier life. Koestler likens this state to what Freud called an 'oceanic feeling', that mystic or religious experience which contrived to escape the closed system of his explanation of the psyche. Paradoxically Koestler argues that whenever he was overtaken by this oceanic feeling the resulting loss of ego caused him to make seemingly rash decisions that radically altered the course of his life for the better. He sees these irrational acts as irruptions from the unconscious, assertions by the deepest portions of the self made for its own good.

The first of these bridge-burning episodes occurs when he burns his Matriculation Book, thereby bringing to a premature close his career as an undergraduate in Vienna. This 'act of apparent lunacy' concealed an underlying need to 'get off the track' of rational determinism and to assert his individuality in a world governed by relativity. (I, pp. 127, 128). The second such incident comes in *The Invisible Writing* when he encourages his assistant at the Ullstein newspaper where he is foreign editor to denounce him as a Communist agent. He knows this will put an end to his career with this ultra-conservative company just as it will bring to an end his usefulness to the Party *Apparat*. Once again he was jumping off the track to preserve his identity. The explanation he gives for his conduct on this occasion also explains the title of the second volume: 'It seems as if . . . one's decisions in these rare moments, however paradoxical or apparently suicidal, followed the commandments of the invisible text, revealed for a split second to the inner self' (II, p. 19). Koestler's life represents a journey from a youth in which he regarded the universe as 'an open book, printed in the language of physical equations and social determinants' to a belief in mid-life that it was actually 'a text written in invisible ink', small fragments of which we contrive to decipher at moments of apparent unreason (II, p. 15). The last instance of such moments comes near the end of the second volume when he describes his resignation from the Communist Party in 1938. That impulsive act was to free him from his full-time commitment to politics and make way for the interests which dominated his later life after completing the autobiography.

Koestler's two autobiographical volumes constitute an extremely complex account of the first portion of his life. This complexity is due to the self-conscious manner in which he exploits the potentialities of the genre. He makes full use of the temporal distinction between his past and present self, refusing to give his complete approval to either version. As he observes of memoirs in *Bricks to Babel*: 'Gains in distance and perspective must be balanced against losses in emotional freshness, for facts are more easily retained than feelings.'[18] On the one hand he inveighs against his own tendency to mock his gauche adolescent self and says he is 'trying to fight this impulse' as he writes about these years (I, p. 76). On the other hand he acknowledges the insight that retrospection has given him into actions, such as his bridge-burning episodes and the periods of hardship that followed them, of which at the time he could make little sense. 'In the rectifying mirror of time, the meagre years which seemed to carry me nowhere appear rich with the fullness of experience; whereas the years of purposeful striving and success were spiritually but a period of marking time' (I, p. 184). Each world seen from the perspective of the other is a looking-glass world in which poverty turns to wealth or vice versa.

Equally sophisticated is Koestler's use of the inner and outer perspectives. He accounts for his six-year spell as a member of the Communist Party and his involvement in the Spanish Civil War by reference both to his life-long guilt complex and the need to assuage it with punishment, and to the prevailing political polarization in central Europe at the time. Inner and outer necessity complement each other, guilt making him obey historical necessity while the latter helps to appease his guilt. Koestler is well aware of the complex, even contradictory nature of the self-portrait he has presented:

The contradictions between sensitivity and callousness, integrity and shadiness, ego-mania and self-sacrifice which appear in every chapter would never add up to a credible character in a novel; but as this is not a novel, he must stand as he was. The seemingly paradoxical can only be resolved by holding the figure against the background of his time, by taking into account both the historian's and the psychologist's approach.
(II, p. 430)

He resists the narrative impulse to simplify, not by abandoning the shape and unity that motivates the imaginative writer, but by forcing it to accommodate contradiction and duality. The idea of a completely integrated personality is a figment of the psychologist's closed system, he argues, which takes no account of the dictates of outer circumstances. On the other hand his infantile experience of fear, anxiety and guilt are co-conspirators, compelling him to plunge into the tides of history under the illusion that he has chosen to do so purely for his own, possibly neurotic, needs. There is no sense of conflict between the two perspectives in Koestler's autobiography. They are two different ways of viewing the same life, each a part-truth in isolation, but in combination a satisfyingly rounded portrait of a particular individual brought up in highly unusual circumstances who found himself swept up in the violent forces of historical change.

As a foreigner his use of English, like Conrad's, lends itself especially well to autobiography because it is so self-conscious. He informs us that he still dreams in Hungarian, German or French. His written English reflects the distance he experiences between instinctual and reflective thought or expression. That puzzling image of 'the rectifying mirror of time' is typical of the way in which he carefully chooses language to express the paradoxical nature of autobiographical retrospection. His brooding, reflective style further adds to that sense of perspectival mutliplicity that also characterizes his attitude to the genre to which he has made such a notable contribution.

Stephen Spender: *World Within World*

Stephen Spender's theories on the nature of autobiography might tempt one to see *World Within World* (1951) as the English equivalent to Koestler's two-volume autobiography. Approximate contemporaries, both writers felt themselves victims of the era into which they were born. 'External things', Spender wrote of the 1930s, 'over which I had no control had usurped my own deepest personal life, so that my inner world became dependent on an outer one.'[19] Like Koestler, Spender argued for the need to balance internal against external versions of the self, subjectivity against objectivity. Yet his practice in his autobiography departs from his theory in showing an obvious preference for the subjective viewpoint wherever subjective and objective come into conflict.

Perhaps this discrepancy between intention and effect is partly a consequence of his writing his autobiography relatively early in life at around the age of forty. In *World Within World* Spender uses a circular form which to some extent diverts attention from the relatively short time span covered by reverting to childhood towards the end of the book. Of the five sections in the book the first gives an achronological sketch of his childhood and youth up to 1928 when he was nineteen. Sections two to four offer a roughly chronological account of his life between nineteen and twenty-nine and relate it to the larger political events of that decade (1928–39). The first half of section five reverts to a personal account of his life as a founder and co-editor of *Horizon* and a fire-fighter during the war. It also very briefly covers the three years between 1947 and 1949 which he spent mostly in the United States where he was intermittently at work on the autobiography. Extracts from it were appearing towards the end of this time in the *Partisan Review*. The final part of section five reverts to his early childhood and schooldays to assert an identity between past and present selves. 'So,' he explains in the Introduction, 'childhood is like wheels within wheels of this book, which begins, and revolves around, and ends with it' (p. viii). Hence the title.

The structure which Spender has chosen reinforces the impression that the subjective perspective is closer to his vision of life than the objective. The objective comes into play fully only in the central portion of the book when politics invaded his private life and forced him to view it from both perspectives during the decade of the thirties. What one learns of his personality during the course of the book shows him to be someone who in his life and his writing cultivated personal feeling and a 'complete submission to experience' (p. 61), that is to personal experience. Auden, after he had known Spender for only a short time, observed that he ought to write, not poetry, but autobiographical prose narrative. In the autobiography Spender accepts Auden's stress on the subjective nature of his literary talent while refusing to confine it to prose: 'As for me, I was an autobiographer restlessly searching for forms in which to express the stages of my development' (p. 138). The same bias towards the personal surfaces in the rather muddled notes Spender made for his own use before writing *World Within World* and subsequently published as 'Confessions and Autobiography'. An ostensible argument in support of the need to relate inner to outer experience in the course of writing an autobiography and vice versa, it none the less ends with a special plea for the unique status of the narrative account of the inner life: 'In a day of pseudo-science when sociologists and psychologists are forever measuring the behaviour of their neighbours, there is a justification for the autobiography which reminds us how lacking in objectivity human beings, who set themselves up as observing instruments, really are.'[20] Like Read and Church, Spender champions subjective autobiography above all for its assertion of individuality and the private interior life.

Spender sees the first forty years of his life as a double curve: originating in the inner life of his childhood but seeking some way of relating it to the world outside, it next seized upon Communism to give it some social reality and purpose; finally it veered back to his childhood belief in the pre-eminence of the inner self and its internal visions. Yet he returns to a subjective perception of life with a difference. As he says of his poetry after he had renounced further direct involvement in politics in the late thirties: 'but I tried to state the condition of the isolated self as the universal condition of all existence' (pp. 254–5). The subjective perspective now strives for objective communication. The circular form he chose to employ for his autobiography mirrors his chronological development which took the middle-aged poet back to his childhood for confirmation of the subjective nature of his talent. His abandonment of his original plan to employ a completely non-chronological form is not, therefore, a failure in boldness on Spender's part, as John Sturrock has suggested.[21] Rather, as a reviewer of the book claimed, his autobiography seeks to apply what he says in it of poetic form ('something which had never been completely achieved' (p. 313)) to his account of his life as a whole.[22] 'Direction', Spender asserts, 'exists in movement, which is living' (p. 313). He perceives both his life and his autobiography as a struggle towards form, an effort 'to create the true tension between . . . inner and outer, subjective and objective, worlds' (p. viii).

This tension between the two perspectives is responsible for Spender's unusual and striking art of portraiture in the book. Nearly all the characters whom he introduces subserve the central self- portrait without losing a vivid sense of their own reality. His father is a good case in point, someone whom his son comes close to caricaturing in his need to fight free of the father's Victorian love of abstractions such as Work, Discipline, Honour or. the Battle of Life. For a poet to be made to think like that is literary suicide. He also attributes to his father his failure at his studies and his need to search out a sexual identity. At one point he reveals a half- conscious wish that his father would die, ostensibly so that he can demonstrate his grief to his father's spirit. When his father does die the seventeen-year-old Spender sees it as a voluntary acknowledgement by his father of his failure as a parent: 'His death seemed a function of our living. We flourished after it' (p. 20). To see his father's life so completely in terms of its significance to his life is the height of subjectivity. Later in the book he is compelled to recognize some of his father's traits in himself – his thirst for fame, for instance, and his love of poetry. Invariably what we learn objectively about the father is always a by-product of his subjective focus on himself and his view of others.

Friends are given similar treatment in the book. They are brought on stage at the point in his life where they answer particular needs in himself at various stages of his development. Auden, for instance, appears on the scene at precisely the moment when Spender needs the confidence to pursue his early attempts at writing poetry. After remarking on what a hold Auden had over his mind at this time considering that he could never follow Auden's philosophical manner of expressing his views, Spender indirectly indicates the way in which subjective and objective approaches to character exist side by side in his portrayal of his friend: 'But it also follows that when I paraphrase his views, I may have modified them, or even reformulated them, out of what I came to understand later. Conversations I quote consist, however, of phrases little altered which have stuck in my memory' (p. 53). Like Koestler's juxtaposition of two closed systems to incorporate the dual perspective, Spender's conjunction of recollected 'fact' and retrospective reinterpretation is his way of combining the Chronicler's urge and the *Ecce Homo* motive. His next mentor is Isherwood on whom he projects all his own need for adventure, independence and excitement. Once again the effect of this subjective projection is to render Isherwood, already an interesting individual in objective terms, even more glamorous as a character than he appears in his own letters from which Spender quotes. After all, how could he hope to rival Spender's imaginary 'Isherwood', a polar explorer in icy northern Europe surrounded by the breakdown in civilization? Even the quotes from Isherwood's letters at this time serve only to show the extent to which Isherwood was playing up to Spender's fantasy of the kind of life he was leading in Berlin. Then there are Spender's various lovers, male and female, who are introduced to illustrate the effect on him of having been brought up in an atmosphere of what he calls 'Puritan decadence'. Walter, Jimmy and Elizabeth all become implicated in his own belief that only sexual guilt restores one to one's own body, a nameless horror to the Puritan decadent. His subjective need to involve his lovers in this sense of guilt, which he feels puts him in touch with reality, is then seen to have wholly objective consequences, such as Jimmy Younger's volunteering to join the International Brigade.

Although Spender claims in his Introduction that only once or twice does the narrative diverge into satire, his use of caricature is more widespread than he suggests. His brother Michael with his mania for competence and efficiency is exhibited like some weird freak at a funfair in order to castigate the rational approach to life which Spender felt threatened his own poetic sensibility. Auden, Isherwood, Humbert Wolfe, J. B. Priestley and a gallery of other minor characters are satirized for similar subjective purposes. Spender's use of what V. S. Pritchett called 'serious farce'[23] in his autobiography might easily have alienated the reader had he not applied it with greatest force to his own portrait. In a long review of *World Within World* Isherwood has discerningly argued that Spender, in reaction against the fool his father had made of himself in his public appearances, 'began . . . by a self-protective instinct, to accentuate those clownish, self-mocking elements in his own personality which would prevent him from . . . turning into a pitiful bore like poor Harold',[24] the father. Spender's use of self-caricature is a sophisticated autobiographical device to counter the unavoidable subjective bias of his self-portraiture. It helps detach the reader from the protagonist and invites him to accept the judgement of a narrator who shows himself able to mock his protagonist's past actions with as much gusto as those of his contemporaries. One such instance of this is the tableau he offers of himself at Oxford reading Blake to the hearties set on breaking up his rooms, and his comment: 'This scene lacked conviction. I was playing the role of the mad Socialist poet, they that of the tough . . .' (p. 34). Again and again he casts himself in a poor part of a badly written play and then as narrator assumes the role of drama critic.

Where satire is applied to others it is normally part of Spender's subjective strategy – condemning their behaviour in order to justify his own adoption of the antithesis. When he turns the satire on himself he is attempting to introduce an objective corrective, although it is normally a corrective to some aspect of his past life to which his present writing self now takes objection. A good instance of this is his justifiably savage caricature of an English novelist and political commissar in Spain who claims, probably incorrectly, to have been responsible for sending a defector in the International Brigade to certain death at the front. Having exposed the pretensions of this writer play-acting at being involved (which he was not) in a war, Spender immediately turns on himself and mocks his childish petulance when not asked to speak at the Writer's Congress in Madrid. He exposes his vanity in posing stock still whenever a camera (actually intended for someone more famous) was pointed in his direction. He rarely spares himself from the criticisms of either himself or others such as Auden or Cyril Connolly. The satire reassures him of his objective existence in a world that doesn't necessarily see him as he sees himself at the time.

Yet his bias towards the subjective is as apparent in his portraiture as elsewhere in the book. As he confesses, 'there was something withdrawn, inaccessible and unexplained . . . about the motives of others' which made him feel confidence only in his ability to reveal the truth of *his* feelings about them (p. 61). So that the objectivity with which he aspires to view all the characters in the autobiography including his own is a mirage in the final analysis, what might be termed 'pseudo-objectivity' compared to his conviction of the reality of his subjective vision. Within the space of an autobiography it is inevitable that the writer should unconsciously settle for a point along the spectrum that represents his own chosen vantage point. It is therefore appropriate that Spender chooses to bring his autobiography to a conclusion by subjective resort to the reiterated imagery of poetry, evoking the sensation he has already described earlier of standing, fire-hose in hand, in the centre of a burning room as 'in the centre of a wheel of my own life where childhood and middle age and death were the same' (p. 322). The achronological form of the book is brought full circle by this image of timelessness which is also an image of the subjectively experienced self.

Storm Jameson: *Journey from the North*

Storm Jameson's impressive two-volume autobiography, *Journey from the North* (1969, 1970), is so starkly personal that one might easily come away with the impression that it uses only an internal perspective. She opens the book by discounting her own significance in public life, despite her presidency of the London centre of P.E.N. during World War II. Even when she turns to her capabilities as an individual she plays them down. She claims to have 'a good but not a great mind',[25] and describes herself as a 'minor writer' (II, p. 325), although the stature of her autobiography alone contradicts this self-estimation. Further she declares from the start that her aim is to 'give a true account of an animal I know only from the inside'. Since 'what one sees from the inside is the seams, the dark tangled roots of feeling and action', she goes on to argue, it may be just as misleading as an externalized self-portrait (I, p. 16). Of course there is still the theoretical presence of an observing self in there somewhere examining the inside as if it were outside. Nevertheless, in keeping with her commitment to the internal perspective she begins by recalling her first memories as a series of key images that she uses as recurrent motifs to structure her account of her life—a field of dazzling white marguerites, the sound of the bell-buoy off Whitby in Yorkshire where she was born and brought up, and the sound of the fishermen's children crying 'Away!' as a ship was cast off. The desire to cast off, for instance, repeatedly surfaces in her life and causes her tremendous disruption and pain which she is none the less unable to avoid, so strongly has the initial pattern imprinted itself on her as a child.

On the other hand you don't have to be a major figure on the stage of world history to see yourself also as the product of forces outside you and to incorporate this social aspect of your personality into any autobiographical account of its development. A child's first experience of interaction between itself and the world comes with its realization that its mother is a separate object from itself. Storm Jameson describes with candour and feeling the extent to which her adult life was conditioned by her complex reaction to her embittered Yorkshire mother. So long as she was an only child her mother was able to take her with her on the voyages which her husband, a sea-captain, was constantly making. These frequent departures before her memory proper begins explain, she says, her life-long restlessness. What stability she had, especially her attachment to Whitby, she claims she owes to her father and to his ancestors who had lived there for hundreds of years. To her mother and all her sea-going ancestors, she writes, she owes her restlessness for which she blames all the subsequent unhappiness in her life. She was told that when she was six months old she set off to climb a dangerously steep staircase in their home and was fetched back and whipped no less than six times. 'Have I', she asks, 'been doing anything ever since except setting off again up those infernal stairs?' (I, p. 28). Out of this external event in her childhood she developed not just her restlessness but the guilt that accompanied it, a guilt which is an internalized form of whipping. Her mother and father soon became estranged and the conflict between their personalities became introjected in their daughter's twin desires for the childhood security of her life in Whitby and that urge to be off anywhere like her maternal ancestors with their 'memories of loss, flight, violence' (I, p. 19).

Storm Jameson's relationship to her mother was a complex combination of identification and rebellion. By leaving her mother when she went to Leeds University she not only began to copy her mother's past restlessness but felt that she had failed her, as she saw it, to satisfy her own needs. Throughout her life she is haunted by a sense of guilt at having abandoned her mother, despite her recognition that 'it is forbidden, on pain of death, to creep back to mother' (p. 51). In abandoning her first husband and – intermittently – her only son, Storm Jameson saw herself merely repeating the pattern established in this first desertion. At one point in the book she even speculates that a psychologist would see her endless departures from people and places and the confusion this caused in her life as the expression of a secret wish to be punished for her errors (I, p. 208). Her catalogue of what she owed to her parents, which comes early in the book, illustrates the way in which she relates the internal view of her make-up to its origins outside herself:

She gave me more than my ludicrous conviction of being responsible for other people, and of being a laughing-stock. I have her bottomless weight of boredom, a never-appeased restlessness, which becomes torture in surroundings I dislike, and the jeering violence I keep out of sight. And, too, a deep, deeper than everything else, indifference – which may only be fear. My working patience and stubbornness I owe to that master mariner with the clouded blue eyes and a mind full of bits and pieces like a tea-chest (I, p. 38).

She both traces the origins of her dominant personality traits back to her parents and shows the effects these have on her social behaviour towards everyone around her. The indifference or failure to love enough which she inherited from her mother is blamed in the autobiography for a whole series of incidents – her neglect of her brother which led to his unhappy apprenticeship with his father's shipping line, her frightened avoidance of either parent on their deathbeds, her failures with men, above all her repeated abandonment of Bill, her only son, to the care of others. This last failure is the cause of some of her most bitter recrimination. She refuses to take refuge in traditional feminist arguments to defend herself from her own accusations. Instead she chooses to live with what she sees as the inevitable guilt that haunts any woman who has attempted to respond to the destructive – because conflicting – demands made on her as a woman and as a writer.

So many of her responses to the outside world are conditioned by her reactions to her mother, the primal embodiment of otherness. Even her involvement in the polarized politics of the thirties is found to have its roots in her early relations with her mother: 'If I looked closely enough, I might see that what made me loathe Fascism was only my hatred of authority, only a mute rebellion against my violently feared and loved mother' (I, p. 295). Like Koestler and Spender, Storm Jameson found herself sucked into the political maelstrom of Europe which radically altered the course of her life and the nature of the books she was writing. Volume I is divided into two parts, the first covering the first forty-one years of her life from 1891 to 1932, the second shorter part concentrating on the 1930s. This structure is intended to reflect the way in which public life and world events took over her life in 1932 when she settled in London again: 'By coming to London when I did, I moved from the margin into the centre at the very moment when the current dragging writers into active politics was gathering force. It was my road to Damascus' (I, p. 293). In the first chapter she offers as one reason for writing her autobiography the fact that as a woman in her seventies her life has spanned three distinct ages, the middle-class heyday before 1914, the interregnum between the two World Wars, and the present age. One would have expected her to reflect this external perspective in the structure of the two volumes. In fact the first part of Volume I spans the pre-1914 stage and half of the interregnum which still held for her the 'illusion of freedom' (I, p. 292). Only the last few years of the interregnum are covered in the second part and these are characterized by disillusion.

Volume II is also divided into two parts, at 1945, but in this case the structure of the book mirrors her conviction that Hiroshima marked an end of an epoch and brought civilization into its final phase in which we all act out our lives against an expectation of global disaster. One reviewer argued that the autobiography reaches its climax at this mid-point, and that the second part of Volume II is a planned anti-climax.[26] In 'The Writer's Situation', an essay Storm Jameson wrote in 1947, she argues that the overriding question of our age is one of survival and that only a great writer is able to tell us about the destiny of man, our destiny, in such a way that we have the courage to live it, and gaily.'[27] The final short paragraph of the first part of Volume II shows her accepting this challenge in her own life and in her autobiographical account of it: 'I had a feeling of exhilaration, even gaiety. Now that none of us is safe, we can really laugh, really mock our pedantic teachers, really live' (II, p. 143). Seen in this light the second part of Volume II represents her attempt to live up to her own highest canons of art. If it proves anti-climactic this is one of the unintended consequences of being a writer with a natural bent towards the internal perspective who is forced to bear witness to external events outside her special purview: 'The world and the frightening things I knew about it distracted me and kept my eyes opened outwards when they might have been sorting the entrails' (II, p. 373).

That Storm Jameson does intend to employ the double perspective is made clear by her choice of epigraph, a quotation from Leonard Woolf's autobiography in which he reiterates the long-held belief that the autobiographer 'should have a double aim: first, to show it and his little ego in relation to the time and place in which he lived his life ... secondly to describe, as simply and clearly as he can, his personal life, his relation, not to history and the universe, but to persons and to himself'.[28] At the same time she is determined to portray the major events of her epoch through a consciously subjective prism. Her first realization that the French were unwilling to stand up to Hitler's expansionism does not originate in some outside source of information; it comes to her in 1937 when she witnesses the adulation with which the German ambassador is received at a Parisian party of wealthy French who fear Socialism and the loss of their riches more than Fascism. Similarly she comes to appreciate the hopeless situation that Czechoslovakia found itself in with Hitler's troops poised on its borders when she hears H. G. Wells tell the Czech ambassador that he will warn President Beneš when he goes to Prague that neither France nor Britain will lift a finger to help his country despite all the friendly speeches he had heard in England. Nazi anti-Semitism takes on reality for her in Budapest where she unwillingly divides her time between a poor Jew and two Jew-haters. The humiliation of Britain's capitulation at Munich becomes a personal embarrassment for her by her need to avoid temporarily her foreign friends living in English exile who make her feel horribly ashamed of her country's desertion of Czechoslovakia. She witnesses at first hand the devastation caused by the war when in 1945 she visits Warsaw, which had been systematically destroyed by the Germans. At the same time she is able to foresee Europe's post-war recovery by encountering the resilient spirit of the young Polish writers she meets there who 'were not simply existing, hanging on blindly, they were living, on the other side of despair, with passion, some at least of them with joy' (II, p. 173). There is never any danger of her autobiography becoming a historical reminiscence because only those fragments of history that were filtered through her immediate experience are given space in the book. Even the position they occupy is dictated by her unflinching concentration on her subjective story.

Later in life she came to regret the amount of time and energy she had spent on her social and public life. Neglected by critics in her later years, she came to feel that she had sacrificed her inner needs for a career in the world of letters: 'There are plenty of people who can go into the world and take part in conversations and the business of a career without losing touch with themselves. Too late, I see I am not one of them, I was only fit to live alone, in society I am worthless' (I, p. 163). She blames herself for having written too many books out of an urge to communicate the life-and-death struggle for civilization she had witnessed in Europe. She sees her prolific production of books as another flight – into words. By 1961 she felt that she had nothing more to write, 'no more words to escape into' (II, p. 370). Within a few months she had started on her autobiography. Much of her fiction is autobiographical. She even made a half-hearted attempt at autobiography in *No Time like the Present* (1933), an account of her immediate ancestors and her own youth up to 1914 which then turns into a description of her opinions at the time. But *Journey to the North* is altogether more ambitious.

Back in 1939 when she was burning personal documents under threat of a German invasion she resolved that if she ever escaped she 'would withdraw from the world and try to write a book, one book which recorded only the essence of my life, the one or two ideas that were mine . . . the one or two feelings, impulses, acts, in which my whole self has been engaged' (II, p. 71). *Journey from the North* is that book. It is an attempt to write a book that 'would change the men and women who read it' (II, p. 371). It was rewritten, she informs us, four or five times. It cultivates a bare style. And it confines its focus to the self. Yet it failed to meet her own high demands, she tells us at the end: 'No one can tell the story of his life. I have tried not to lie, and doubt I have told more lies than truth. I could have written it in several ways, all half true, half a lie' (II, p. 382). Yet the very process of making overt the intrinsic paradox of the genre helps make of her 'failure' an outstanding contribution to the genre, a uniquely personalized account of the impact on the internal life of external events.

By employing a double perspective all four of these writers have sought to check the natural tendency of the genre towards exaggerating the uniqueness of an individual's internal life. As Charles Rycroft has put it, 'The process of detaching that thread which is one's own life from the fabric which has been simultaneously woven by those around one introduces an inherent bias towards egocentricity, at the expense of objectivity, and towards exaggeration of one's difference and alienation from others.'[29] Probably the bias is especially strong among literary autobiographers since creative writers tend to see themselves as critics of their society and times and champions of individual freedom. No amount of self-caricature or use of the quoted opinions of others about the protagonist can overcome the way in which the genre serves to reveal yet more about the inner personality of the autobiographer by the manner in which he is attempting to counter his subjective perspective. Nevertheless the genre has another compensatory characteristic which it shares with all written texts. This is the way the text assumes an impersonal life of its own. So the autobiographical act simultaneously transforms the life it is describing and externalizes it by embodying it in a form that, because of its employment in the past, imposes conditions and creates expectations outside the writer's control. The author of an autobiographical text finds himself cast in the role of a narrator of a scenario over which he has only partial control presenting and commenting on an even more remote protagonist whose script and actions appear to be largely out of his control. It is the meeting of the writer's ego with the impersonal requirements of the genre which gives to those autobiographies that recognize this the fascination of an encounter between life and art, personal and impersonal, internal and external perspectives. These generic complexities have been recognized by its best practitioners since Augustine's time. But the realization and exploitation of the full potentialities of this fascinating form of narrative has become far more common in the course of this century, which is why it is so rich in examples of autobiographies that are likely to remain classics of their kind. They deserve to be recognized as works of art quite as worthy of attention and as rewarding as those better known examples of imaginative writing, novels, plays and poems.

A Personal Postscript

When I read through this book after an interval of time I was struck by the fact that it was both less theoretical than I had originally intended and yet paradoxically less subjective than I had wanted it to be as well. I could argue that this simply reflects the richness of the subject which lends itself to a whole variety of treatments. More pertinent is the question why I opted for the relatively objective, yet empirical approach.

This book originated in my experience of teaching adults literature as part of my university's extra-mural provision. During the period leading up to the writing of this book I was becoming converted to the idea that literature could be taught successfully only if the experience of reading a text could be related to the students' own experiences of analogous feelings and situations. The teaching of literature, like the reading of literature, is not what for a long time I presumed it to be – a discipline of subordinating one's subjective responses to an 'objective' reading of the text, an attempt to hear the author's voice unmuffled by the clamour of the reader's ego. A text assumes significance for a reader only if it can enter into the reader's current life and interact with his or her anxieties, memories, hopes, fantasies, desires and the like. So that to read or to teach a text entails finding a point of contact between what the writer is actually saying (in so far as this can be ascertained) and whatever elements of personal experience the text brings to life in the reader.

Granted this approach to the reading of literature, autobiography naturally has an advantage over other forms of imaginative writing, especially in a teaching situation. All of us feel powerful currents of feeling when, for instance, reading a well-written account of someone's childhood, because we all constantly retell to ourselves the story of our own childhood to accommodate our evolving adult needs and circumstances. So I found myself quite naturally making increasing use of autobiographical texts in courses in order to encourage students to seek a genuine, particularized point of contact between their lives and their experience of reading literature. In doing so I repeatedly met in students an initial insistence on questioning how truthful any autobiographical text was. This insistence normally proved a major impediment in reaching any such point of contact. Simultaneously the same average reader has a tendency to judge an autobiography on the basis of its author's moral conduct. For instance, one of the great innovators of the genre, Rousseau, attracted considerable opprobrium from a group of my students on the conflicting grounds that he had got his facts wrong and that he had owned up to some despicable deeds.

If it is thought that this reaction is somewhat naive, rather than simply typical, I would refer the reader to a recent book of generic criticism, *The Art of Autobiography in 19th and 20th Century England* by A. O. J. Cockshut published in 1984. He shows a similar tendency to judge the texts he is considering by appraising the behaviour of their authors. In a chapter on childhood autobiography Cockshut criticizes the autobiographical accounts of both W. H. Hudson and Forrest Reid for failing to achieve what he feels Edwin Muir succeeded in doing – bringing into harmony the solitary paradisial experience of childhood with the world and values of an adult social being. Words charged with moral values reminiscent of F. R. Leavis like 'reconciliation' and 'wholeness' are repeatedly evoked as standards against which Hudson's and Forrest Reid's narratives are found wanting. Cockshut, in other words, judges the texts as if he were judging the protagonists in them, and he further alleges that the adult authors are still suffering the moral defects of their earlier selves.

This tempting slide from text to author, I found, occurs more readily and commonly among readers of autobiography than among readers of fiction schooled in the 'discipline' of reading ever since the arrival of the New Criticism. This attitude to autobiography is partly an offshoot of the way most critics and autobiographers still tend to pillage a writer's autobiography for the 'facts' about his or her life. Despite the excellent theoretical work produced over the past two decades by Continental and American critics, autobiography is still largely treated by the literary establishment as a reservoir of information, which, to be sure, needs to be used with caution, but which nevertheless is felt primarily to constitute a repository of personally vouched-for insights into the author's earlier life. Only within the small circle of critics of the genre is it now a commonplace that an autobiography is likely to throw more light on the normally ageing autobiographer than on the earlier self about whom the book is ostensibly written.

Since I undertook this book to fill a yawning gap and so to make more possible the teaching as well as the reading of twentieth-century British autobiography, I have had no hesitation in addressing it principally to the average reader and student of autobiography rather than to the more sophisticated theorists of the genre. I felt it necessary, as I do in any class of students new to the study of autobiography, first of all to lead the reader of this book through the arguments and experience necessary to establish the independence of autobiography from historical-cum-biographical truth. By concentrating the attention of the reader of autobiography instead on the autobiographer's honesty to his current conception of the shape his past has taken, I felt I could simultaneously eliminate the desire to pass moral judgement on him. The reader's interest refocuses – away from the behaviour of the protagonist and on to the portrait that gradually emerges of the author who is narrating his own story. The autobiographer's search through his past for an understanding of the self that is conducting the search provides the theme for the second half of this book. Since the search is conducted in written form it is as a narrative that it has to be judged, differing only in its generic intention from so-called 'imaginative' narratives which at any rate can at times proximate to historical accuracy more consistently than it does.

If the product of this reasoning is a book which appears to its more sophisticated readers to take too little for granted I can only repeat that all my experience with widely read, well-educated students is that widespread misunderstanding of the genre prevails. Further, most of the classic autobiographies considered in this book are little read, and even more rarely read as works of literature in their own right. I have had to assume acquaintance with the major texts selected for analysis in this book. But I have resolutely aimed it at the student and reader relatively new to a study of the genre. I realize that in attempting to pre-empt the tendency among readers of autobiography to pass moral judgements I have perhaps adopted an unnecessarily impersonal, judicious tone of voice. In fact the rich variety of autobiographical texts currently in print elicits from me an equally wide variety of responses. But, as I wrote in the Introduction, I have for the purposes of this book used each text purely as an example of one of its many facets. I have done this in the name of clarity.

Yet life isn't as clear as it appears in this book. My own life during the writing of the book entered a period of complexity and obscurity which has made me vividly aware of how little justice I have been able to do in the preceding pages to the conflicting currents of thought and feeling, of text and subtext, conscious intention and unconscious revelation which account for much of the fascination which the genre holds for me. If I say, what I feel to be true, that were I to start afresh I would write a very different book now, I am saying no more than what almost every autobiographer considered in this book has demonstrated in the course of refashioning his past in the distorting mirror of his present self-image.

Notes

All publishers are London-based unless specified otherwise.

Introduction

1 Viz. Wayne Shumaker, *English Autobiography: Its Emergence, Materials and Forms*, Berkeley and Los Angeles, University of California Press, 1954.
2 Bernard Shaw, *Sixteen Self Sketches*, Constable, 1949, p. 6.
3 Pamela Hansford Johnson, *Important to Me*, New York, Scribner's, 1974, p. 9.
4 *The Complete Works of Montaigne*, ed. Donald Frame, California, Stanford University Press, 1957, p. 504.
5 William C. Spengemann, *The Forms of Autobiography: Episodes in the History of a Literary Genre*, New Haven, Yale University Press, 1980, p. 168.
6 E. Stuart Bates, *Inside Out: An Introduction to Autobiography*, Oxford, Blackwell, 1936, p. 2.
7 Motlu Konuk Blasing, *The Art of Life: Studies in American Autobiographical Literature*, Austin, University of Texas Press, 1977, p. xi.
8 Baron Richard A. Butler, *The Difficult Art of Autobiography*, Oxford, Clarendon Press, 1968, p. 19.
9 Roy Pascal, *Design and Truth in Autobiography*, Routledge & Kegan Paul, 1960, p. 5.
10 See my article on 'Sexual Identity in Modern British Autobiography', *Prose Studies*, 7 (Summer 1985).
11 André Maurois, *Aspects of Biography*, trans. S. C. Roberts, Cambridge, Cambridge University Press, 1929, p. 111.

1 Factual Accounts of the Self

1 Dean Ebner, *Autobiography in Seventeenth Century England: Theology and the Self*, The Hague, Mouton, 1971.

2 Somerset Maugham, *The Summing Up*, Pan Books, 1976, p. 10.

3 *Aspects of Biography*, p. 159.

4 George Orwell, *Collected Essays, Journalism, and Letters*, ed. S. Orwell and I. Angus, Vol. 3, Secker & Warburg, 1968, p. 156.

5 Sigmund Freud, *Letters of Sigmund Freud*, ed. Ernst Freud, Hogarth Press, 1961, p. 391.

6 *Sixteen Self Sketches*, p. 42.

7 Gerald Brenan *A Life of One's Own*, Cambridge, Cambridge University Press, 1979, p. xii.

8 W. H. Hudson, *Far Away and Long Ago*, Dent, 1939, p. 2.

9 Georges Gusdorf, 'Conditions and Limits of Autobiography', trans. J. Olney, in *Autobiography. Essays Theoretical and Critical*, ed. J. Olney, Princeton, New Jersey, Princeton University Press, 1980, p. 42.

10 John Wain, *Sprightly Running*, Macmillan, 1963, p. 233.

11 W. H. Davies, *The Autobiography of a Super-Tramp*, Oxford, Oxford University Press, 1980, p. 15. All page references are to this edition.

12 Cf. R. J. Stonesifer, *W. H. Davies: A Critical Biography*, Cape, 1963, pp. 40–1.

13 Ibid., p. 45.

14 Cf. S. Gwynn, 'The Making of a Poet', *The Living Age*, 265 (May 1910) p. 492; J. Freeman, *Herman Melville*, New York, Macmillan, 1926, p. 87; and G. B. Shaw's Preface to *The Autobiography of a Super-Tramp*, p. 10.

15 *W. H. Davies: A Critical Biography*, p. 120.

16 W. H. Davies, *The Adventures of Johnny Walker, Tramp*, Cape, 1926, p. 8.

17 W. H. Davies, *New Statesman and Nation*, 18 March 1933, pp. 338–40.

18 *Collected Essays, Journalism and Letters*, Vol. 1, p. 114.

19 George Orwell, *The Road to Wigan Pier*, Harmondsworth, Middlesex, Penguin Books, 1962, p. 133.

20 Mabel Furze speaking in a BBC *Omnibus* programme of 1970, cited in Bernard Crick, *George Orwell: A Life*, Harmondsworth, Middlesex, Penguin Books, 1982, p. 198.

21 George Orwell, *Down and Out in Paris and London*, Secker & Warburg, 1951, p. 9. All page references are to this edition.

22 *Collected Essays, Journalism and Letters*, Vol. 1, p. 114.

23 M[uriel] H[arris], *Manchester Guardian*, 9 January 1933.

24 S. Greenblatt, 'Orwell as Satirist', *George Orwell: A Collection of Critical Essays*, ed. R. Williams, Englewood Cliffs, New Jersey, Prentice-Hall, 1974, p. 105.

25 *George Orwell: A Life*, p. 200.

2 Internal Verification

1   *Design and Truth in Autobiography*, p. 72.

2   *Aspects of Biography*, p. 142.

3   G. K. Chesterton, *Autobiography*, Hutchinson, 1936, pp. 35–6.

4   Eugine Donato, 'The Ruins of Memory: Archaeological Fragments and Textual Artefacts', *Modern Language Notes* 93 (May 1978) p. 576. Donato's use of a quotation from Freud gives a misleading impression of what were Freud's considered views on the operation of memory.

5   A. E. Coppard, *It's me, O Lord!*, Methuen, 1957, p. 97.

6   Henry Green, *Pack My Bag*, Hogarth Press, 1940; new edn, 1979, pp. 7–8 and 11–12.

7   *It's me, O Lord!*, p. 9.

8   Wyndham Lewis, *Rude Assignment*, Hutchinson, 1950, p. 103.

9   Georg Misch, *A History of Autobiography in Antiquity*, trans. E. W. Dicks, Routledge & Kegan Paul, 1950, Vol. 1, p. 11.

10 *Design and Truth in Autobiography*, p. 191.

11 Geoffrey Grigson, *The Crest on the Silver: An Autobiography*, Cresset Press, 1950, p. 27.

12 J. N. Morris, *Versions of the Self*, New York, Basic Books, 1966, p. 213.

13 Frank Harris, *My Life and Loves*, ed. John F. Gallagher, W. H. Allen, 1964. All page references are to this edition.

14 Quoted in Philippa Pullar, *Frank Harris*, Hamish Hamilton, 1975, p. 356.

15 Letter to Hesketh Pearson, 2 May 1927, ibid., p. 110.

16 Ibid., p. 18.

17 Christopher Isherwood, *Christopher and His Kind*, New York, Farrar, Straus & Giroux, 1976, p. 60.

18 *Frank Harris*, p. 15.

19 Clive James, *Unreliable Memoirs*, Cape, 1980, pp. 11–12.

20 George Moore, *Hail and Farewell: Ave, Salve, Vale*, ed. Richard Cave, Colin Smythe, 1976. All page references are to this edition.

21 Quoted by F. S. Lyons in 'Goodbye to Dublin', *Times Literary Supplement*, 4 February 1977, p. 122.

22 Joseph Hone, *The Life of George Moore*, Gollancz, 1936, pp. 200–1.

3   Fact and Fiction

1   Anna R. Burr, *The Autobiography: A Critical and Comparative Study*, Constable and Houghton Mifflin, 1909, p. 148.
2   Autobiography in Seventeeth Century England, p. 155.
3   Cf. *The Beginnings of Autobiography in England*, a paper delivered by James M. Osborn at the fifth Clark Library seminar, 8 August 1959, which cites Thomas Whythorne's autobiography written *c*. 1576.
4   Phyllis Grosskurth, 'Where was Rousseau?' *Approaches to Victorian Autobiography*, ed. George P. Landow, Athens, Ohio, Ohio University Press, 1979.
5   Cf. 'The Study of Autobiography: A Bibliographical Essay' in Spengemann, *The Forms of Autobiography: Episodes in the History of a Literary Genre*, New Haven, Conn., Yale University Press, 1979.
6   François Mauriac, *Journal II*, Paris, La Table Ronde, 1937, p. 138.
7   *It's me, O Lord!*, p. 9.
8   Ibid., p. 247.
9   Virginia Woolf, *Collected Essays*, Vol. IV, Hogarth Press, 1966, pp. 221–8.
10 J. D. Beresford, *Writing Aloud*, Collins, 1928, p. 80.
11 Ibid., p. 3.
12 Ibid., p. 201.
13 Michel Leiris, *L'âge d'homme*, Paris, Librairie Gallimard, 1939; *Manhood*, trans. Richard Howard, Cape, 1968, pp. 13, 15.
14 Alan Sillitoe, *Raw Material*, W. H. Allen, 1972, rev. edns 1974, 1978. All page references are to the 1978 edition.
15 *It's me, O Lord!*, p. 216.
16 *English Autobiography*, p. 111.
17 The *Art of Life: Studies in American Autobiographical Literature*, p. xiii.
18 H. G. Wells, *Experiment in Autobiography*, Gollancz, 1934; reissued by Faber & Faber, 1984, p. 22.
19 *The Crest on the Silver*, pp. 3–4.
20 Richard Aldington, *Life for Life's Sake*, New York, Viking Press, 1941, p. 404.
21 Elizabeth Bowen, *Seven Winters*, Longman's Green, 1943, p. 48.
22 *L'âge d'homme*, p. 20.
23 *A History of Autobiography in Antiquity*, p. 11.
24 Cf. Frederick R. Karl, *Joseph Conrad: The Three Lives*, Faber & Faber, 1979, pp. 658, 661, 669.
25 Letter from Conrad to F. M. Hueffer, 31 July 1909, quoted in Jocelyn Baines, *Joseph Conrad: A Critical Biography*, Harmondsworth, Middlesex, Penguin Books, 1971, p. 420.

26 Cf. *Joseph Conrad: The Three Lives*, pp. 655, 662.

27 *Joseph Conrad: A Critical Biography*, p. 420.

28 Joseph Conrad, *A Personal Record: Some Reminiscences*, Gresham Publishing Co., 1925, p. xix. All page references are to this edition.

29 Ford Madox Ford, *Thus to Revisit*, Chapman & Hall, 1921, pp. 53, 55.

30 Ibid., p. 44.

31 *Design and Truth in Autobiography*, p. 177.

32 *Writing Aloud*, p. 49.

33 E. M. Forster, *Aspects of the Novel*, Harmondsworth, Middlesex, Penguin Books, 1962, p. 52.

34 David Thorburn, *Conrad's Romanticism*, New Haven, Conn., Yale University Press, 1974, p. 64.

35 *Aspects of the Novel*, p. 70.

36 Stephen Spender, *World Within World*, Hamish Hamilton, 1951, p. ix.

37 John Cowper Powys, *Autobiography*, Macdonald, 1934, reissued 1967, p. 642.

38 Christopher Isherwood, *Lions and Shadows*, New English Library, 1974, p. 5. All page references are to this edition.

39 Letter from Isherwood to A. Clodd, 7 December 1953, quoted in Brian Finney, *Christopher Isherwood: A Critical Biography*, Faber & Faber, 1979, p. 130.

40 Christopher Isherwood, 'Autobiography of an Individual', *The Twentieth Century*, 149 (May 1951) p. 405.

41 Cf. Stephen Spender, 'On Being a Ghost in Isherwood's Berlin', *Mademoiselle*, 79 (September 1974) p. 139.

42 Alan Wilde, *Christopher Isherwood*, New York, Twayne Publishers, 1971, p. 20.

43 Evelyn Waugh, 'Author in Search of a Formula', *Spectator*, 160 (25 March 1938), p. 538.

44 Interview between the author and Isherwood, 3 September 1976.

45 Jon Bradshaw, 'Reflections of an Anglo-Californian', *Vogue*, 131 (December 1974), p. 84.

4   The Comic Perspective

1 Molière, *Oevres complètes*, ed. R. Jouanny, Paris, Garnier, 1962, Vol. I, p. 632.

2 Roland Barthes, *S/Z*, Paris, Seuil, 1970, p. 17.

3 Michael Sprinker, 'The Fictions of the Self', in Olney (ed.), *Autobiography: Essays Theoretical and Critical*, pp. 326, 327.

4 *A Sort of Life*, p. 7.

5   *Experiment in Autobiography*, p. 22.
6   Sigmund Freud, *The Complete Psychological Works*, trans. James Strachey, Hogarth Press and The Institute of Psychoanalysis, 1953–74, Vol. VIII, p. 206.
7   Martin Grotjahn, *Beyond Laughter*, New York, McGraw Hill, 1957, p. 259.
8   Ludwig Jekels, 'On the Psychology of Comedy', *Theories of Comedy*, ed. Paul Lauter, Garden City, New York, Doubleday, Anchor, 1964, p. 431.
9   John Mortimer, *Clinging to the Wreckage: A Part of Life*, Weidenfeld & Nicolson, 1982, p. 14. All page references are to this edition.
10 John Mortimer interviewed by Sheridan Morley, 'British Business at the Bar', *The Times*, 4 January 1982, p. 9.
11 J. L. Borges, *A Personal Anthology*, Pan Books, 1972, pp. 171–2.
12 'British Business at the Bar'.
13 *Theories of Comedy*, p. 452.
14 Northrop Frye, *A Natural Perspective*, New York, Columbia University Press, 1965, p. 121.
15 *Theories of Comedy*, p. 455.
16 *The Letters of J. R. Ackerley*, ed. Neville Braybrooke, Duckworth, 1975, p. 256.
17 Ibid., p. 304.
18 J. R. Ackerley, *My Father and Myself*, New York, Harcourt Brace Jovanovich, 1969, p. 7. All page references are to this edition.
19 *My Sister and Myself: The Diaries of J. R. Ackerley*, ed. Francis King, Hutchinson, 1982, p. 135.
20 Ibid., pp. 25–6.
21 *The Complete Psychological Works*, Vol. VIII, p. 222.
22 Northrop Frye, *Anatomy of Criticism*, Princeton, New Jersey, Princeton University Press, 1957, p. 180.
23 V. S. Pritchett, *Autobiography*, English Association, June 1977, p. 4.
24 *Anatomy of Criticism*, p. 168.
25 Robert M. Torrance, *The Comic Hero*, Cambridge, Mass., Harvard University Press, 1978, especially pp. 1–11.
26 V. S. Pritchett, *A Cab at the Door/Midnight Oil*, Harmondsworth, Middlesex, Penguin Books, 1979, p. 56. All page references are to this edition.
27 *Anatomy of Criticism*, p. 169.
28 *Why Do I Write? An Exchange of Views between Elizabeth Bowen, Graham Greene and V. S. Pritchett*, ed. E. Bowen, Percival Marshall, 1948, p. 14.
29 *Autobiography*, pp. 9–10.

30 V. S. Pritchett, *Mr Beluncle*, Chatto & Windus, 1951, p. 104.
31 Graham Hough, 'The True Truth about Himself', *New Statesman*, 75 (23 February 1968), p. 240.
32 *Why Do I Write?*, p. 18.
33 *Anatomy of Criticism*, pp. 169–70.

5 Childhood

1 *A History of Autobiography in Antiquity*, p. 8.
2 Roland Barthes, 'To Write: An Intransitive Verb?', *The Structuralists: From Marx to Levi-Strauss*, ed. R. and F. De George, Garden City, New York, Doubleday, 1972, p. 163.
3 James Olney, *Metaphors of the Self: The Meaning of Autobiography*, Princeton, New Jersey, Princeton University Press, 1972, pp. 29–30.
4 *The Complete Psychological Works*, Vol. IX, p. 150.
5 Luann Walthar, 'The Invention of Childhood in Victorian Autobiography', *Approaches to Victorian Autobiography*, p. 64.
6 *Autobiography*, p. 54.
7 Leonard Woolf, *An Autobiography*, Oxford University Press, Vol. I, 1980, p. 10.
8 Edwin Muir, *An Autobiography*, Hogarth Press, 1980, p. 125.
9 Ibid., p. 281.
10 Herbert Read, *The Contrary Experience: Autobiographies*, Faber & Faber and New York, Horizon Press, 1963, p. 20. All page references are to this edition.
11 *The Collected Works of C. G. Jung*, trans. R. F. C. Hull, Routledge & Kegan Paul, 1953–71, Vol. 8, paras 750, 751.
12 Ibid., Vol. 9, Part II, para. 40.
13 Laurie Lee, 'Writing Autobiography', *I Can't Stay Long*, André Deutsch, 1975, p. 52.
14 Ibid., p. 51.
15 Ibid., p. 49.
16 Ibid., pp. 49–50.
17 Ibid., p. 53.
18 Laurie Lee, *Cider With Rosie*, Hogarth Press, 1973, p. 9. All page references are to this edition.
19 *The Collected Works of C. G. Jung*, Vol. 8, para. 98.
20 *I Can't Stay Long*, p. 44.
21 Edward Edinger, *Ego and Archetype*, Harmondsworth, Middlesex, Penguin Books, 1973, pp. 3–26.
22 *The Collected Works of C. G. Jung*, Vol. 17, para. 331a.
23 Ibid., Vol. 10, para. 315.

24   Laurie Lee, *As I Walked Out One Midsummer Morning*,
     Harmondsworth, Middlesex, Penguin Books, 1971, p. 48.
25   *The Collected Works of C. G. Jung*, Vol. 10, para. 315.
26   Richard Church, 'The Art of Autobiography', *The Cornhill
     Magazine*, 171 (Winter 1960–1), p. 475.
27   Ibid., p. 469.
28   Norman O. Brown, *Life Against Death*, Middletown, Conn.,
     Wesleyan University Press, 1959, p. 31.
29   Richard Church, *The Voyage Home*, Heinemann, 1964, p. 165.
30   Richard Church, *Over the Bridge*, Heinemann Educational Books,
     1966, p. 24. All page references are to this edition.
31   *The Voyage Home*, p. 238.
32   'The Art of Autobiography', p. 475.
33   *The Collected Works of C. G. Jung*, Vol. 15, para. 103.
34   Ibid., para. 100.
35   Ibid., para. 103.

6   Parents and Children

1   *The Childhood of Edward Thomas: A Fragment of
     Autobiography*, Faber & Faber, 1938; new edn, 1983, p. 13.
2   Sergei Aksakov, *Years of Childhood*, trans. J. D. Duff, Oxford,
     Oxford University Press, 1983, p. 2.
3   *The Complete Psychological Works*, Vol. XIII, p. 243.
4   Jean Rhys, *Smile Please: An Unfinished Autobiography*, André
     Deutsch, 1979, p. 46.
5   Edmund Gosse, *The Life of Philip Henry Gosse*, London, 1890,
     quoted in Evan Charteris, *The Life and Letters of Sir Edmund
     Gosse*, Heinemann, 1931, pp. 222–3.
6   *The Life and Letters of Sir Edmund Gosse*, p. 308.
7   Edmund Gosse, *Father and Son*, ed. P. Abbs, Harmondsworth,
     Middlesex, Penguin Books, 1983. All page references are to this
     edition.
8   *The Life and Letters of Sir Edmund Gosse*, p. 311.
9   Ibid., p. 307.
10   David Grylls, *Guardians and Angels*, Faber & Faber, 1978, p. 177.
11   *The Life and Letters of Sir Edmund Gosse*, p. 8.
12   *The Complete Psychological Works*, Vol. XIII, p. 145.
13   Ibid., Vol. XXI, p. 30.
14   Ibid., p. 74.
15   Ibid., p. 125.
16   Roger J. Porter, 'Edmund Gosse's *Father and Son*: Between Form
     and Flexibility', *Journal of Narrative Technique*, 5 (September
     1975), p. 181.

17 Cf. André Gide, *Journals 1889–1949*, trans. J. O'Brien, Harmondsworth, Middlesex, Penguin Books, 1967, pp. 253–5.

18 Patricia Beer, *Mrs Beer's House*, Hutchinson, 1978, p. 212.

19 Ibid., p. 156.

20 Ibid., p. 101.

21 *The Letters of W. B. Yeats*, ed. A. Wade, Rupert Hart-Davis, 1954, p. 589.

22 R. Ellmann, *Yeats: The Man and the Masks*, Faber & Faber, 1961, p. 2.

23 Joseph Ronsley, *Yeats's Autobiography*, Cambridge, Mass., Harvard University Press, 1968, p. 21.

24 *The Letters of W. B. Yeats*, p. 589.

25 Ibid., p. 389.

26 W. B. Yeats to Lily Yeats, 29 December 1914, cited in William Murphy, *Prodigal Father: The Life of John Butler Yeats (1839–1922)*, Ithaca, New York, Cornell University Press, 1978, p. 423.

27 *The Letters of W. B. Yeats*, p. 606.

28 *The Collected Works of C. G. Jung*, Vol. 4, para. 693.

29 Ibid., para. 739.

30 Ibid., para. 744.

31 J. B. Yeats to Susan Yeats, 1 November 1872, cited in Joseph Hone, *W. B. Yeats 1865–1939*, Harmondsworth, Middlesex, Penguin Books, 1971, pp. 17–18.

32 W. B. Yeats, *Autobiographies*, Macmillan, 1955, p. 27. All page references are to this edition.

33 *The Letters of W. B. Yeats*, pp. 602–3.

34 W. B. Yeats, *Memoirs*, ed. D. Donoghue, Macmillan, 1972, p. 19.

35 *The Collected Works of C. G. Jung*, Vol. 4, para. 729.

36 Osbert Sitwell, *Left Hand, Right Hand!*, Reprint Society, 1946, p. viii. All page references are to this edition.

37 Stephen Spender, *The Making of a Poem*, Hamish Hamilton, 1955, p. 66.

38 *The Collected Works of C. G. Jung*, Vol. 16, para. 978.

39 Ibid., para. 976.

40 Osbert Sitwell, *Great Morning*, Macmillan, 1948, p. 152.

41 Ibid., p. 41.

42 Osbert Sitwell, *Laughter in the Next Room*, Reprint Society, 1950, p. 90.

43 Osbert Sitwell, *The Scarlet Tree*, Reprint Society, 1947, p. 262.

44 Cf. John Pearson, *Facades: Edith, Osbert and Sacheverell Sitwell*, Fontana, 1980, p. 51.

45 *The Collected Works of C. G. Jung*, Vol. 15, para. 4.

7 Autobiography as Self-Analysis

1 *The Letters of D. H. Lawrence*, ed. G. T. Zytaruk and J. T. Boulton, Cambridge, Cambridge University Press, 1981, p. 90.
2 John Lehmann, *The Whispering Gallery*, Longmans Green, 1955, p. vii.
3 *Autobiographical Acts*, p. 164.
4 *Unreliable Memoirs*, p. 11.
5 *The Complete Psychological Works*, Vol. XXII, p. 234.
6 *The Collected Works of C. G. Jung*, Vol. 4, para. 449.
7 *Experiment in Autobiography*, p. 22.
8 Edmund Blunden, *Undertones of War*, Harmondsworth, Middlesex, Penguin Books, 1937, p. 8.
9 Robert Graves, *Goodbye to All That*, Harmondsworth, Middlesex, Penguin Books, 1960, p. 240. All page references are to this edition.
10 Robert Graves, *Good-bye to All That*, Cape, 1929, p. 443.
11 Ibid., p. 439.
12 Robert Graves, *But It Still Goes On*, Cape, 1930, p. 7.
13 Ibid., p. 13.
14 Paul Fussell, *The Great War and Modern Memory*, New York, Oxford University Press, 1975, p. 208.
15 Robert Graves, 'The Art of Poetry XI', *Paris Review*, 47 (1968–9), p. 135.
16 *But It Still Goes On*, p. 42.
17 *The Great War and Modern Memory*, p. 207.
18 *Good-bye to All That* (1929), p. 437.
19 Martin Seymour-Smith, *Robert Graves. His Life and Works*, Hutchinson, 1982, p. 12.
20 *Good-bye to All That* (1929), p. 440.
21 Ibid., p. 13.
22 Siegfried Sassoon, *Memoirs of an Infantry Officer*, Faber & Faber, 1965, p. 108. All page references are to this edition.
23 Dame F. Corrigan, ed., *Siegfried Sassoon: Poet's Pilgrimage*, Gollancz, 1973, p. 84.
24 *The Great War and Modern Memory*, p. 104.
25 Siegfried Sassoon, *Sherston's Progress*, Faber & Faber, 1936, p. 242. All page references are to this edition.
26 Siegfried Sassoon, *Siegfried's Journey 1916–1920*, Faber & Faber, 1945, p. 55. All page references are to this edition.
27 Siegfried Sassoon, *Memoirs of a Fox-Hunting Man*, Faber & Faber, 1960, p. 61. All page references are to this edition.
28 R. Hart-Davis, ed., *Siegfried Sassoon Diaries 1915–1918*, Faber & Faber, 1983, p. 22. All page references are to this edition.

29 *The Great War and Modern Memory*, p. 100.

30 John Lucas, 'For ever England', *London Review of Books*, 5 (16 June – 6 July 1983), p. 19.

31 Siegfried Sassoon, *The Old Century and Seven More Years*, Faber & Faber, 1938, p. 128.

32 Caroline Moorhead, 'Osborne, still tearing into the troglodytes', *The Times*, 10 October 1981, p. 6.

33 John Osborne, *A Better Class of Person: An Autobiography 1929–1956*, Faber & Faber, 1981, p. 272. All page references are to this edition.

34 Moorhead, 'Osborne, still tearing into the troglodytes'.

35 Ibid.

36 John Osborne, 'Face to Face', *The Playwrights Speak*, ed. W. Wager, Longmans, 1969, p. 76.

37 Jeremy Treglown, 'He who plays the prince', *Times Literary Supplement*, 16 October 1981, p. 1190.

38 Ibid.

39 *The Collected Works of C. G. Jung*, Vol. 17, para. 328.

40 Ibid., para. 291.

8 Myths and Dreams in Autobiography

1 Thomas De Quincey, *Suspira de Profundis*, Macdonald, 1956, p. 448.

2 *The Complete Psychological Works*, Vol. V, p. 608.

3 Ibid., Vol. IV, p. 160.

4 Ibid., Vol. IX, p. 152.

5 Olney, *Autobiography: Essays Theoretical and Critical*, p. 338.

6 *The Complete Psychological Works*, Vol. V, p. 525.

7 C. G. Jung, *Memories, Dreams, Reflections*, Collins/Fontana, 1967, pp. 17–19.

8 Ibid., p. 30.

9 *Christopher Isherwood: A Critical Autobiography*, p. 277.

10 Ibid., p. 279.

11 Forrest Reid, *Apostate*, Constable, 1926, p. 4. All page references are to this edition.

12 Brian Taylor, *The Green Avenue*, Cambridge, Cambridge University Press, 1980, p. 110.

13 Forrest Reid, *Private Road*, Faber & Faber, 1940, p. 12.

14 *The Green Avenue*, p. 4.

15 *Private Road*, p. 11.

16 *The Green Avenue*, pp. 108–9.

17 Ibid., pp. 108, 115.

18 Ibid., p. 26.

19   *Private Road*, p. 197.
20   *The Collected Works of C. G. Jung*, Vol. 10, para. 325.
21   *The Green Avenue*, pp. 24–6.
22   Mary Bryan, *Forrest Reid*, Boston, Mass., Twayne Publishers,
     G. K. Hall, 1976, p. 23.
23   *The Collected Works of C. G. Jung*, Vol. 8, para. 505.
24   Edwin Muir, *An Autobiography*, Hogarth Press, 1980, p. 49. All
     page references are to this edition.
25   Cf. Willa Muir, *Belonging: A Memoir*, Hogarth Press, 1968,
     p. 19.
26   Quoted in P. H. Butter, *Edwin Muir: Man and Poet*, Oliver &
     Boyd, 1966, p. 159.
27   P. H. Butter, ed., *Selected Letters of Edwin Muir*, Hogarth Press,
     1974, p. 116.
28   Ibid., pp. 111–12.
29   'Manuscript Notes for An Autobiography', quoted in Butler,
     Edwin Muir: Man and Poet, p. 247.
30   Ibid., p. 248.
31   E. W. Mellown, *Edwin Muir*, Boston, G. K. Hall, Twayne
     Publishers, 1979, p. 87.
32   *Selected Letters of Edwin Muir*, p. 101.
33   Edwin Muir, *The Story and the Fable*, Harrap, 1940, p. 193.
34   Edwin Muir, 'Yesterday's Mirror: Afterthoughts to an
     Autobiography', *Scots Magazine* (New Series), XXXIII
     (September 1940), pp. 404–10.
35   Edwin Muir, *Collected Poems 1921–1958*, Faber & Faber, 1960,
     p. 141.
36   Cf. *Selected Letters of Edwin Muir*, p. 182.
37   *The Collected Works of C. G. Jung*, Vol. 10, para. 304f.
38   *The Story and the Fable*, p. 263.
39   Genesis 4: 19.
40   *The Story and the Fable*, p. 261.

9   The Spiritual History of the Self

1   *English Autobiography*, p. 74.
2   *Versions of the Self*, p. 91.
3   *Autobiography*, p. 81.
4   C. S. Lewis, *Surprised by Joy*, Geoffrey Bliss, 1955, p. 15.
5   Ibid., p. 219.
6   *The Collected Works of C. G. Jung*, Vol. 11, para. 168.
7   *Autobiographies*, pp. 115–16.
8   Cf. *Design and Truth in Autobiography*, p. 52.

9 W. H. Hudson, *Far Away and Long Ago*, Dent: Everyman's
   Library, 1939, 1967, p. 196. All page references are to the 1967
   edition.
10 W. H. Hudson, *Idle Days in Patagonia*, Dent, 1893, 1954,
   p. 205.
11 Ibid., p. 206.
12 Ibid., p. 205.
13 Ibid.
14 Jocelyn Brooke, *The Orchid Trilogy*, Harmondsworth,
   Middlesex, Penguin Books, 1981, p. 47.
15 Nicholas Shakespeare, 'From Pampas to Paddington', *Times
   Literary Supplement*, 26 November 1982, p. 1292.
16 *Idle Days in Patagonia*, pp. 110–11.
17 Ibid., p. 75.
18 Ibid., p. 213.
19 John Cowper Powys, *Autobiography*, Macdonald, 1967, p. xx.
   All page references are to this edition.
20 John Cowper Powys, Llewelyn Powys, *Confessions of Two
   Brothers*, Sinclair Browne, 1982, pp. 38–9.
21 Cf. Belinda Humphrey, ed., *Essays on John Cowper Powys*,
   Cardiff, University of Wales Press, 1972, p. 342.
22 Ibid., Introduction.
23 Ibid., p. 339.
24 Ibid., p. 127.
25 Ibid., p. 339.
26 Louis Marlow, *Welsh Ambassadors (Powys Lives and Letters)*,
   Chapman & Hall, 1936, p. 27.
27 John Cowper Powys, *A Philosophy of Solitude*, Cape, 1933,
   pp. 82–3.
28 Jeremy Hooker, *John Cowper Powys*, Cardiff, University of
   Wales Press, 1973, p. 10.

10 The Double Perspective: The Self and History

1 Erik H. Erikson, 'Gandhi's Autobiography: The Leader as
   Child', *The American Scholar*, 35 (1966) pp. 632–46.
2 *Design and Truth in Autobiography*, p. 185.
3 Ibid., pp. 180–1.
4 *Father and Son*, p. 33.
5 Ibid., p. 160.
6 Vera Brittain, *Testament of Youth*, Fontana, 1979, p. 12.
7 *Experiment in Autobiography*, p. 15.
8 Cf. Norman and Jeanne Mackenzie, *The Time-Traveller: The
   Life of H. G. Wells*, Weidenfeld & Nicolson, 1973, p. 386.

9   Cf. H. G. Wells, 'My Auto-Obituary', 1943, reprinted in *H. G.
     Wells: Interviews and Recollections*, ed. J. R. Hammond,
     Macmillan, 1980, pp. 117, 119.

10   Arthur Koestler, *Dialogue with Death*, Collins with Hamish
     Hamilton, 1954, p. 97.

11   Arthur Koestler, *Scum of the Earth*, Macmillan, New York,
     1941, pp. 263–4.

12   Arthur Koestler, *Bricks to Babel*, Hutchinson, 1980, p. 315.

13   Arthur Koestler, *The Trail of the Dinosaur*, Collins, 1955,
     Preface.

14   Cf. Harold Harris, 'Author', *Encounter*, 61 (July–August 1983),
     p. 25. Cf. Arthur and Cynthia Koestler, *Stranger on the Square*,
     Hutchinson, 1984.

15   Arthur Koestler, *The Invisible Writing*, Collins with Hamish
     Hamilton, 1954, II, p. 423. All page references preceded by 'II'
     are to this edition.

16   Cf. Pascal, *Design and Truth in Autobiography*, pp. 148–50,
     181.

17   Arthur Koestler, *Arrow in the Blue*, Macmillan, New York,
     1952, I, p. 245. All page references preceded by 'I' are to this
     edition.

18   *Bricks to Babel*, p. 315.

19   Stephen Spender, *World Within World*, Hamish Hamilton, 1951,
     p. 190. All page references are to this edition.

20   Stephen Spender, *The Making of a Poem*, p. 67.

21   Cf. John Sturrock, 'The New Model Autobiographer', *New
     Literary History*, 9 (Autumn 1977), pp. 51–63.

22   Cf. *Times Literary Supplement*, 13 April 1951, p. 228.

23   V. S. Pritchett, Review of *World Within World*, *New Statesman*,
     41 (14 April 1951), p. 427.

24   Christopher Isherwood, 'Autobiography of an Individualist', *The
     Twentieth Century*, 149 (May 1951), p. 407.

25   Storm Jameson, *Journey from the North*, Collins and Harvill
     Press, Vol. I: 1969, Vol. II: 1970, I, p. 15. All page references
     preceded by 'I' or 'II' are to this two-volume edition.

26   'Mother of literary exiles', *Times Literary Supplement*, 12 March
     1970, p. 272.

27   Storm Jameson, *The Writer's Situation*, Macmillan, 1950, p. 36.

28   Leonard Woolf, *Downhill All The Way*, Hogarth Press/Reader's
     Union, 1968, p. 40.

29   Charles Rycroft, 'Viewpoint: Analysis and the Autobiography',
     Times Literary Supplement, 27 May 1983, p. 541.

Bibliography

Book-length General Studies in English of Autobiography

All publishers are London-based unless specified otherwise.

Bates, E. Stuart, *Inside Out: An Introduction to Autobiography*, Oxford, OUP, 1936

Blasing, Motlu Konuk, *The Art of Life: Studies in American Autobiographical Literature*, Austin, Texas, University of Texas Press, 1977

Bottrall, Margaret, *Every Man a Phoenix: Studies in Seventeenth Century Autobiography*, Chester Springs, Pennsylvania, Dufour, 1958

Brignano, Russell C., *Black Americans in Autobiography: An Annotated Bibliography*, Durham, Duke University Press, 1974

Briscoe, Marie Louise, ed., *A Bibliography of American Autobiography 1945–1980*, Madison, Wisconsin, University of Wisconsin Press, 1982

Bruss, Elizabeth W., *Autobiographical Acts: The Changing Situation of a Literary Genre*, Baltimore, Johns Hopkins University Press, 1976

Buckley, Jerome Hamilton, *The Turning Key: Autobiography and the Subjective Impulse Since 1800*, Cambridge, Massachusetts, Harvard University Press, 1984

Burr, Anna Robeson, *The Autobiography: A Critical and Comparative Study*, Boston, Houghton Mifflin, 1909

Butler, Lord Richard A., *The Difficult Art of Autobiography*, Oxford, Clarendon Press, 1968

Butterfield, Stephen, *Black Autobiography in America*. Amherst, University of Massachusetts Press, 1974

Clark, A. M., *Autobiography: Its Genesis and Phases*, Edinburgh, Oliver and Boyd, 1935

Cockshut, A. O. J., *The Art of Autobiography in 19th and 20th Century England*, New Haven, Yale University Press, 1984

Coe, Richard N., *When the Grass was Taller: Autobiography and the Experience of Childhood*, New Haven, Yale University Press, 1984

Cooley, Thomas, *Educated Lives: The Rise of Modern Autobiography in America*, Columbus, Ohio State University Press, 1976

Couser, G. Thomas, *American Autobiography: The Prophetic Mode*, Amherst, University of Massachusetts Press, 1979

Delany, Paul, *British Autobiography in the Seventeenth Century*, Routledge & Kegan Paul, 1969

Eakin, Paul J., *Fictions in Autobiography*, Princeton, Princeton University Press, 1985

Earle, William, *The Autobiographical Consciousness: A Philosophical Inquiry into Existence*, Chicago, Quadrangle Books, 1972

Ebner, Dean, *Autobiography in Seventeenth-Century England: Theology and the Self*, The Hague, Mouton, 1971

Egan, Susannah, *Patterns of Experience in Autobiography*, Chapel Hill, University of North Carolina Press, 1984

Fleishman, Avrom, *Figures of Autobiography: The Language of Self-Writing*, Berkeley, University of California Press, 1983

Genre (special number), 6 (March and June 1973)

Georgia Review (special number on autobiography and biography), Summer 1981

Kaplan, Louis, *et al.*, *A Bibliography of American Autobiographies*, Madison, University of Wisconsin Press, 1961

Landow, George P. ed., *Approaches to Victorian Autobiography*, Athens, Ohio, Ohio University Press, 1979

Lillard, Richard G., *American Life in Autobiography: A Descriptive Guide.* Stanford, Stanford University Press, 1956

Matthews, William, *British Autobiographies: An Annotated Bibliography of British Autobiographies Published or Written before 1951*, Berkeley, University of California Press, 1955

Mehlman, Jeffrey, *A Structural Study of Autobiography: Proust, Leiris, Sartre, Levi-Strauss*, Ithaca, Cornell University Press, 1974

Misch, Georg, *A History of Autobiography in Antiquity*, trans. E. W. Dickes, 2 vols, Cambridge, Massachusetts, Harvard University Press, 1951

Modern Language Notes (special number on autobiography) 93 (May 1978)

Morris, John N., *Versions of the Self: Studies in English Autobiography from John Bunyan to John Stuart Mill*, New York, Basic Books, 1966

New Literary History (special number on autobiography) 9 (Autumn 1977)

Olney, James, *Metaphors of Self: The Meaning of Autobiography*, Princeton, Princeton University Press, 1972

Olney, James, ed., *Autobiography: Essays Theoretical and Critical*, Princeton, Princeton University Press, 1980

Osborn, James M., *The Beginnings of Autobiography in England*, Los Angeles, University of California Press, 1959

Padover, Saul K., *Confessions and Self-Portraits: 4600 Years of Autobiography*, New York, J. Day & Co., (1957)

Pascal, Roy, *Design and Truth in Autobiography*, Routledge & Kegan Paul, 1960

Pearson, Hesketh, *Ventillations: Being Autobiographical Asides*,
   Philadelphia, Pennsylvania, 1930
Peyre, Henri, *Literature and Sincerity*, New Haven, Yale University Press,
   1963
Pilling, John, *Autobiography and Imagination*, Routledge & Kegan Paul, 1981
Porter, Roger J., and H. R. Wolf, *Voice Within: Reading and Writing
   Autobiography*, New York, Knopf, 1973
Pritchett, Sir Victor S., *Autobiography* (presidential address), English
   Association, 1977
Salaman, Esther P., *The Great Confession: From Aksakov and De Quincey
   to Tolstoy and Proust*, Allen Lane, 1973
Sayre, Robert F., *The Examined Self: Benjamin Franklin, Henry Adams,
   Henry James*, Princeton, Princeton University Press, 1964
Sewannee Review (special number on biography and autobiography) 85
   (Spring 1977)
Shea, Daniel B., *Spiritual Autobiography in Early America*, Princeton,
   Princeton University Press, 1968
Shumaker, Wayne, *English Autobiography: Its Emergence, Materials and
   Forms*, Berkeley and Los Angeles, University of California Press, 1954
Smith, Sidonie A., *Where I'm Bound: Patterns of Slavery and Freedom in
   Black American Autobiography*, Westport, Connecticut, Greenwood
   Press, 1974
Spacks, Patricia M., *Imagining a Self: Autobiography and Novel in
   Eighteenth-Century England*, Cambridge, Massachusetts, Harvard
   University Press, 1976
Spengemann, William C., *The Forms of Autobiography: Episodes in the
   History of a Literary Genre*, New Haven, Connecticut, Yale University
   Press, 1979
Stone, Albert E., *Autobiographical Occasions and Original Acts: Versions
   of Identity from Henry Adams to Nate Shaw*, Philadelphia,
   Pennsylvania: University of Pennsylvania Press, 1982
Taylor, J. Lionel, *The Writing of Autobiography and Biography*, Hull,
   privately printed, 1926
Webber, Joan, *The Eloquent 'I': Style and Self in Seventeenth-Century
   Prose*, Madison, University of Wisconsin Press, 1968
Weinstein, Arnold, *Fictions of the Self: 1550–1800*. Princeton, Princeton
   University Press, 1983
Weintraub, Karl J., *The Value of the Individual: Self and Circumstance in
   Autobiography*, Chicago, University of Chicago Press, 1978
Wethered, H. N., *The Curious Art of Autobiography from Benvenuto
   Cellini to Rudyard Kipling*, Christopher Johnson, 1956

See also: *Biography* (an interdisciplinary quarterly devoted to biography
   and autobiography) Honolulu, University Press of Hawaii, 1978–

Index